CONCEPTS AND MEASUREMENT OF
QUALITY OF LIFE IN HEALTH CARE

Philosophy and Medicine 47

CONCEPTS AND MEASUREMENT OF QUALITY OF LIFE IN HEALTH CARE

Edited by

LENNART NORDENFELT

Department of Health and Society,
Linköping University,
Linköping, Sweden

KLUWER ACADEMIC PUBLISHERS

DORDRECHT / BOSTON / LONDON

A C.I.P. Catalogue record for this book is available from the Library of Congress.

ISBN 0-7923-2824-8

Published by Kluwer Academic Publishers,
P.O. Box 17, 3300 AA Dordrecht, The Netherlands.

Kluwer Academic Publishers incorporates
the publishing programmes of
D. Reidel, Martinus Nijhoff, Dr W. Junk and MTP Press.

Sold and distributed in the U.S.A. and Canada
by Kluwer Academic Publishers,
101 Philip Drive, Norwell, MA 02061, U.S.A.

In all other countries, sold and distributed
by Kluwer Academic Publishers Group,
P.O. Box 322, 3300 AH Dordrecht, The Netherlands.

Printed on acid-free paper

Printed in the Netherlands

TABLE OF CONTENTS

SECTION III / MEASURING QUALITY OF LIFE IN HEALTH CARE

EDITOR'S PREFACE

Questions concerning the notion of quality of life, its definition, and its applications for purposes of assessment and measurement in social and medical contexts, have been widely discussed in Scandinavia during the last ten years. To a great extent this discussion mirrors the international development in the area. Several methods for the assessment and measurement of quality of life have been borrowed from the UK and the US and then further developed in northern Europe. But there has also been an internal development. This holds in particular for the social arena, where Scandinavia has had a special tradition both in theory and practice.

In this volume an attempt is made to illustrate some aspects of the philosophical, and in general theoretical, discussion concerning quality of life in Scandinavia. In addition, some prominent scholars from other parts of Europe, i.e., France, the Netherlands, the UK and Italy, have been invited to contribute. The volume is divided into three sections. The first contains philosophical analyses of the general notion of quality of life and proposes a number of different explications. The second section considers various applications of the notion of quality of life in health care. The papers serve to disentangle some intellectual and ethical problems that stem from these applications. The third section is more practical and focuses on methods of measuring quality of life in medicine and health care.

A few of the authors represented here presented their essays at a symposium on Quality of Life in Linköping, Sweden, in April 1991. The others were invited after that date to write original contributions for this volume.

As the editor of this volume, I wish to express my gratitude to a number of persons for their help in preparing the book. I wish particularly to thank Professor Stuart Spicker and Mr. Malcolm Forbes for their hard work in transforming the papers into a publishable form. I also wish to thank my colleague, Henk ten Have for his good advice and for his willingness to accept this manuscript in the European subseries for studies in Philosophy of

Medicine and Health Care. Finally, my gratitude is extended to Monika Thörnell and Lena Hector at my own Department, who have spent many hours transforming the texts at various stages of completion to the final camera-ready manuscripts.

Linköping, January 1994

Lennart Nordenfelt

INTRODUCTION

QUALITY OF LIFE AS A NEW GOAL FOR MEDICINE AND HEALTH CARE

There is a growing concern among members of the health care system, as well as among the public, that the ultimate goal of medicine and health care cannot be simply the cure of disease and the forestalling of death. Several factors in the development of medicine, as well as in the panorama of health problems, have contributed to this concern.

A particularly important factor is the technological progress within medicine. The technology in, for instance, intensive care units has made it possible to save or at least to prolong many lives which would previously have terminated. But the life that has been prolonged can, as in many cases of severe cancer, be a life of great pain and disability. Or it can, as in the case of late stages of Alzheimer's disease, be a life which is devoid of human dignity.

Another factor has to do with the changing nature of the disease panorama. The dominant diseases are now not acute, in general treatable, infectious diseases but instead chronic diseases and chronic impairments for which there is at present no chance of effective cure. In these cases health care must aim at a goal which is distinct from the elimination of disease. It must aim at improving the life of the patient in other respects; it must support, encourage, and in general provide patients with the means to cope with a life that involves a serious long-term health problem.

A third factor is more theoretical and ideological. I have in mind the criticism of the so-called machine model of the human being. This is the model which is supposed to form the basis of much scientific medicine in its concentration on the human being as a biological organism, and its lack of interest in the human being as a social agent. Really effective and humane medicine, the critic emphasizes, must understand and care about a person as an integrated, feeling, and active being. It is the quality of such an integrated person's life that we should care about, not primarily the person as a biological organism.

As a result of these developments, there is now in medicine an extensive literature dealing with quality-of-life (QoL) issues. In some of the clinical specialities, such as cancer care, cardiovascular care and psychiatric care, attempts have been made, at least partially, to define therapeutic success in terms of the patient's QoL. For this purpose a number of instruments or scales, mainly of a questionnaire type, have been devised. Some of these scales have been frequently used and are known under such names as *The Nottingham Health Profile* and *The Sickness Impact Profile* [13].

But the crucial question now is: what do these instruments in fact measure? What is in fact defined as an important goal of medicine by these instruments? What concept of QoL do they presuppose? And how is this notion related to health, in particular when the latter is not just taken to be identical with the absence of disease? These are some of the questions to be dealt with in this volume. In this introductory essay I shall in particular highlight the first, most basic question. I shall discuss various platforms for analyzing and defining the notion of QoL. I shall try to show that the various possible definitions will have profoundly different implications for the procedure of applying the notion in the context of health care.

TOWARDS A CONCEPTUAL ANALYSIS

The notation "QoL" is composed of two terms: "quality" and "life." Both are in need of analysis. Notwithstanding the increasingly rich literature on QoL, the basic notion of life has been almost completely neglected in this context. (For an outstanding exception see [20].) Though I shall not here attempt a full-length analysis, some conceptual observations are necessary for the present purpose.

An important distinction is one between a *complete life* and a *partial life*. Moreover, there are at least two dimensions along which the degree of completeness can be measured. One dimension has to do with *time*; another has to do with the totality of *aspects of life*. A complete life in the former sense is composed of the continuous series of life-events that a particular person goes through during his or her existence from birth to death. A complete life in the latter sense is the sum total of all the aspects of his or her existence at a certain moment or during a certain period of time. A *maximally complete* life is, then, the sum total of all the aspects of a person's existence during his or her entire life-time.

The idea of an *aspect of life* is very essential. Since it is easily seen that apart from God no one is going to study every aspect of a person's life, we must make some selection, and preferably a well-motivated one. The selection should be guided by the particular purpose that an assessment or measurement of life-quality has.

At least the following main aspects of life could be considered.

a. The experiential aspect of life; the sum total of a person's sensations, perceptions, cognitions, emotions, and moods.
b. The activities in life; the sum total of a person's actions.
c. The achievements in life; the sum total of the results of a person's actions.
d. Events in a person's life, those that the subject is aware of or which are for other reasons ascribed to him or her, or both.
e. Circumstances surrounding a person, either those that the subject is aware of or which are for other reasons ascribed to him or her, or both.

Moreover, these five categories can be mixed into combinations so that we receive a great, if not infinite, number of interpretations of the notion of life. One interpretation is the maximal one. A maximal life contains all the mentioned elements, i.e., it is the sum total of a person's experiences, activities, and achievements, as well as all events and circumstances ascribable to him or her.

It is evident, as illustrated by the essays in this volume, that different theories of life-quality have focused on different aspects of *life* in the full sense of the word. This partly has to do with what aspect the theorist judges to be an important aspect, one that is worthwhile considering for his or her particular purpose. Unfortunately, however, the choice of aspects is not always clearly motivated. (For a discussion of some cases, see my analysis in [18].)

It is very important to draw the distinction between, on the one hand, aspects of a person's life which could be objectively (or at least intersubjectively) ascribed to him or her, and, on the other hand, such aspects as the subject perceives or by which he or she is causally influenced. This is one of three senses in which we can talk about *objective* quality of life and *subjective* quality of life.

A person can be placed in a set of circumstances, or he or she may have a set of personal properties, which for one reason or another have a high value. But it may be the case that the person is not aware of them. And it may even be the case, although this is more unlikely, that the circumstances or the personal properties do not causally influence the person's conscious life. Let me illustrate from the field of welfare and well-being. A man may have a great sum of money in his bank account of which he is not aware. In some sense, the "objective" sense, this man is a wealthy man and can receive a high score on some welfare scale. On the other hand, if the man is not aware of the money, he cannot make use of it for any purpose. The subjective well-being is not affected.

Similarly, we can imagine that a person receiving some medical treatment has obtained bodily and mental resources that could be used for many good purposes. On an ability and energy scale this person scores high. On the other hand, if the person is not aware of this fact, and it does not influence his or her life, then this does not affect his or her subjective well-being.

As a corollary to this we can note the following interpretations of the term 'welfare'. On the one hand, it may signify a state of affairs generally taken to be advantageous to anyone in such a situation. On the other hand, it can signify a state which is perceived and appreciated by a particular person, or the person's conscious life may be causally influenced by it. In the latter sense, then, 'welfare' becomes a completely relational term.

A second distinction marked by the terms 'objective' and 'subjective' is the one between non-mental states of affairs and mental ones. A state of affairs, or an event external to a person's experiential life, is often called "objective" (because it is accessible to other people than the subject him- or herself), whereas a mental event, be it perceptive, cognitive, or emotional, is called "subjective", according to this interpretation. (These distinctions are well elaborated in Musschenga's essay [15].)

What, then, about the notion of *quality*? In ordinary and philosophical discourse the term 'quality' may be used in both a neutral and a normative sense. In the former case 'quality' is simply coextensive with 'property', as in the locution: "What qualities does this substance have?" In the QoL discourse, however, 'quality' is almost exclusively used in the second, normative sense, indicating some kind of *evaluation* of a phenomenon, a ranking of the phenomenon along an evaluative scale. It is essential in the QoL-context to be able to say that a person's life-quality is high or low or, even clearer, that the quality is good or bad.

We are then faced with two difficult issues in need of exploration. The first concerns finding an evaluative dimension along which to rank life; the second concerns who should evaluate a person's life?

(1) Which is (or which are) the evaluative dimension(s) along which a person's life (or certain aspects of this person's life) should be assessed or measured?

There are many values to be considered. There are, for instance, moral values, prudential values, aesthetic values, intellectual values, values of humour, values of decency, values of welfare as well as values of experiential well-being. A total life or any part of such a total life can be evaluated along all of these value-dimensions.

Furthermore, there may be an evaluation *among* the dimensions. One dimension may be considered to have a higher dignity or be preferable for some purpose than another. In talking about QoL (unqualified) one may thus mean the quality according to what one takes to be the *most important* evaluative dimension. We can here suggest an analogy with the Platonic universe of ideas. All ideas are valuable entities, but the idea of goodness is the most valuable of all.

(2) Who should be the evaluator of a particular person's life? This might then also entail: Who is to choose the dimension to be utilized in the evaluation?

A Platonic-Aristotelian answer to these two questions would run along the following lines. There are a number of dimensions of value; there is also a scale among the values, where one dimension of value can be compared to another dimension of value. These evaluations are given once and for all, in the same sense as mathematical truths are given once and for all. There is in principle no room for personal choice among values. As a result, it does not matter who exercizes an evaluation in a particular case. As long as this person has the proper insight into the realm of absolute values, then this person will also come out with *the* correct evaluation of this person's life-quality. In particular, Aristotle gave us a substantial treatment of the perfect life and he suggested that the best life, all things considered, is the life in accordance with the highest of virtues. Such a life is what he called a *eudaimonian* life. (For a detailed analysis, see Ostenfeld [22].)

There is no single contemporary view on this matter. There are contemporary views which come close to the Aristotelian one (consider, for instance, von Wright's on the good life [25] and Nussbaum's treatment of non-relative virtues [21]. But there are also contemporary views which are the direct opposite of an Aristotelian approach (see Nordenfelt's [19] and Sandøe's essays [24]. For a critical analysis of this approach see Cattorini and Mordacci [5]). Moreover, there are views which lie somewhere between (Kajandi's [10]). The anti-Aristotelian view entails the conclusion that there is no hierarchy of values given once and for all. All values are chosen by individuals and therefore every assessment of life-quality is dependent on the person who makes the evaluation.

This is the problematic starting-point for the *empirical science* of quality-of-life research. What dimensions of evaluation and what criteria should we follow? There seem to be two kinds of plausible strategies available — one more collective and paternalistic, the other more liberal and individualistic.

The former strategy, which could be realized in a variety of ways, would entail the following. A number of "experts" or politicians come together to decide what is the essence of QoL. They decide first what aspects of life are the most important for the particular purpose chosen — for instance, circumstantial aspects, activity aspects, or experiential aspects. Second, through a consensus-reasoning or through a simple majority vote they decide on the scale along which individual lives should be measured.

This could be done in a more or less *a priori* fashion. The people involved in the decision could have an Aristotelian view and attempt to work out the details of their *eudaimonia*-concept, and then try to establish this as the basis for assessment. (An example of this is, perhaps, the need-approach for the evaluation of QoL which is to be further described below.) A more *a posteriori* (or democratic) procedure would involve making an empirical investigation and obtaining an idea about how people in general evaluate their lives. This is the procedure chosen for the establishment of some measurement scales in health care; see, e.g.,[8] [23].

The *individualistic* strategy entails that there is no general instrument for the assessment of QoL, or at least there is no presupposition that such an instrument is available. The person who has the task of making the assessment, whether this is a physician or a social worker, permits the subjects themselves to make the evaluation. The subjects can thus make their life-evaluation according to their own preferences. *X's quality of life becomes identical with X's own evaluation of his or her QoL.*

We have here identified the third sense of the phrase "subjective quality of life". I.e., the person's subjective quality of life is identical with his or her own evaluation of the life.

Furthermore, the subject's own evaluation can be performed on two levels, both of which may have practical significance. The subject can provide an evaluation in a very radical sense of the word. He or she can decide what values should be considered in a particular assessment of QoL. The subject can, for instance, say the following: "I am immobile and in great pain, but I prefer this state because it makes me morally sensitive. Since my primary goal is to be a moral person, I prefer this state of affairs to a state where I could move about freely and have less pain."

The other interpretation is the more conventional. According to this the subject is asked to provide an evaluation of his or her QoL concerning a certain well-defined sector of aspects, and within a certain realm of values.

But what, now, is the deeper meaning of the locution "X's evaluation of his or her QoL"? And how can we come to understand this evaluation? It seems as if the locution can be interpreted in two ways, one more superficial than the other, and these interpretations have methodological consequences, moreover. The most straightforward and superficial interpretation is: X's evaluation of X's QoL is identical with X's explicit answer to a question concerning X's QoL, presupposing that X has correctly understood the question and attempts to answer honestly.

The other interpretation is the following: X's evaluation of his or her life is the evaluation that X would make given that X could organize his or her value system coherently and could give a proper assessment of all relevant details of his or her life (according to this system).

This, then, is an *ideal* notion of subjective evaluation of QoL. No subject can fully accomplish this, however. Most of us lack coherent value-systems and we are unable to remember and truly grasp all relevant facts about our lives. Moreover, some individuals are particularly poor at this task. The persons may not have the intellectual qualifications. Some human beings, not only infants and the mentally retarded, do not understand the notion of a value. Moreover, they may have little possibility of grasping a broader view of their lives, let alone of organizing their value-system coherently.

Thus, to achieve the deep subjective evaluation of a life seems to be an extremely difficult task. On the other hand, the more subtle interpretation appears to be the more intellectually reasonable. A person's evaluation of his or her life cannot, strictly speaking, be identical with what he or she "honestly" says in reply to a question. If it were, there would be no room for a subject to make a mistake in assessing his or her QoL. (Questions concerning honesty and the possibility of self-deception in relation to QoL are in the focus of Siri Naess' contribution to this volume [17].)

But if nobody can achieve full insight into his or her own quality of life, even on this individualistic interpretation, what are the methodological implications? Here some theorists seem to propose a compromise among a number of methods.

First, it is said, there is good reason to believe that a normal adult is capable of providing a rough estimate of his or her QoL. The person may, however, need some help from an external assessor to more thoroughly think about all relevant areas and all relevant events. As in the case of psychotherapy, the assessor can help the subject to come to describe his or her value system, and to help the subject think about all relevant matters in order to obtain a reasonably accurate account of his or her QoL.

Some people — infants, the mentally retarded, some psychotics and the truly senile — have no or very little capacity to express orally their view about QoL. In such cases other people must assist and complete the evaluation for them. But this can still, and should (according to the individualistic philosophy) be done *from the subject's point of view*. It remains the subject's QoL that should be assessed and no-one else's. (A discussion of some of these issues is found in Kajandi's and Musschenga's essays [10] [15].)

What, then, are the methods for such an other-person evaluation? Here we should be able to rely both on common sense and scientific, i.e., psychological insights. There are, it may be argued, many ways in which we can determine if a person prefers the situation in which he or she is placed and in general likes to live. In particular, a mother can know quite well if her baby is well and happy. The baby reveals it with its body language, with cries and laughter, through other facial expressions and signs of ease and tension, etc. The other-person interpretation, in this case, remains, however, an interpretation of the subject's signs, although these signs may not be conventionally linguistic.

Let me recapitulate: The fact that another person takes part in the evaluation of a subject's QoL need not entail the abandonment of the individualistic (subjectivistic) notion of QoL. To be sure, the other-person interpretation is difficult in practice and there is a significant risk of mistakes. But this, I think, is an unavoidable state of affairs.

ON THE CONTEMPORARY DISCUSSION OF QOL

Quality of life and health

Thus far my approach has been rather abstract. I have sketched a few metaphysical platforms for making QoL-assessments. But how could this reasoning be applied to today's QoL-discussion and the actual measurements which are made in, for instance, social work and health care? Let me, in answering this, focus on the field of *health care* as do most of the other contributions to this volume. I shall in this discussion also stress the crucial role of the *purpose* with which a particular QoL-measurement is undertaken. (A background to today's QoL-discussion in health care is given in Bury's essay [4].)

A first question concerns the scope of the notion of life in QoL-measurements in the health-care context. It is clear that the idea of an entire life in the temporal sense is excluded. The typical assessment of life-quality is supposed to serve a practical purpose; among other things it is presumed to serve *the purpose of improving one or more persons' QoL*. Thus the assessment cannot in such cases presuppose that the subject has already lived his or her whole life (i.e., has died). Moreover, the task is to assess and measure something which is *at present* the case, i.e., such relatively stable factors in life as define the person's present life-situation.

But certainly not all aspects of the present are to be considered. Today's QoL assessors tend to select very carefully among aspects, mainly concentrating on personal — bodily and mental — properties as well as the person's experiential aspects. Sometimes a set of circumstantial facts are also included. (For a discussion concerning the selection of aspects, see Björk and Roos [2].)

There is a tendency, then, in contemporary assessments of people's QoL, apart from the extreme cases of metaphysical or religious assessments, to concentrate on the present, and also to concentrate on such features of the present judged to be relatively permanent for or "surrounding" the individual. "Permanence" here essentially means that the factors judged are not likely to improve by themselves; indeed, they may be more likely to "deteriorate" by themselves.

In order to determine what scales of evaluation should be used, it is important to distinguish between the QoL-instruments used in psychiatry and social medicine, and those employed in somatic medicine. The former instruments tend to cover more aspects of human life. They are principally used as guides in clinical work and to improve the QoL of individual patients. The somatic ones, on the other hand, can be very limited in content and may only mirror aspects which are closely related to the subject's *health*. These instruments may be used both in the clinical context and in large-scale assessments of medical technologies, which are undertaken for a purpose like resource allocation. (For a comprehensive account of different applications of QoL-instruments, see Fitzpatrick and Albrecht's essay [7].)

As an example of an instrument from psychiatry, I mention only one that is developed by a Swedish researcher, Madis Kajandi [9], which is based on a concept developed by Siri Naess [16]. The elements of Kajandi's instrument are the following: *external life conditions*, (which can be divided into working life, personal economy, and housing quality); *interpersonal relations* (divided into partner-relations, relations of friendship, relations to parents and relations to one's children); *inner mental states* (divided into involvement in life, energy, self-actualization, freedom, self-assuredness, self-acceptance, emotional experiences, security, and general mood).

It is evident that Kajandi's instrument ranges across a multitude of aspects of human life: the most stable circumstances in the individual's life, most of his or her personal relations and all important mental states, as well as many of his or her mental abilities. But what evaluation scale is used?

Kajandi, like most theorists in this area, says rather little about the nature of the scale. It seems, however, to be presupposed in these contexts that the scale of evaluation has to do with *the individual's welfare or well-being*. One important aspect, if not completely exclusive, is whether the *subject likes his or her life, or not*. This is not unexpected. People seek psychiatric care because they frequently have personal problems of a basic kind. They are depressed, they feel anguish, they cannot cope with their lives. Their most urgent wish is to transcend these problems and improve their well-being. There is, under these circumstances, very little room for an evaluation of their lives from a moral, aesthetic, or intellectual point of view, unless these aspects happen to have a significant influence on the subject's well-being.

In *somatic medicine* the focus is much more precise. This is so because of the often quite limited and practical purpose of the QoL-measurement. Here, for example, the purpose can be the testing of a new medical methodology, e.g., a drug which affects a very limited bodily function, or which has only a pain-relieving effect.

The question can then, of course, be raised: What are we measuring? Are we simply measuring health, or are we measuring something "between" health and QoL, e.g., subjective health? (Note that some leading textbooks refer to many of the instruments in this *somatic* area as "health-measuring instruments" [13].)

There is here conceptual unclarity which is in need of further analysis. Indeed, there are fundamental differences in our conceptions of health. Some theorists, who align themselves more closely with a bio-statistical analysis of health, represented by the work of C. Boorse [3], view health as the absence of disease, and consider diseases as virtually measurable along physiological parameters.

According to this view, the use of a questionnaire which seeks to determine how the subject feels (e.g., if he or she has any pain, or if he or she can cope with stairs), entails the measurement of something over and above somatic health. On this view, then, it is justifiable to claim that the measurement concerns QoL.

On the other hand, assume that we adopt a holistic conception of health (reflected in many contemporary texts) entailing that a person is healthy if and only if, he or she is capable of realizing certain vital goals in life [18]. A holistic concept of this kind automatically infers that the absence of pain, and the basic ability to cope with stairs, are contained in the notion of health itself. QoL for a holistic health theorist must entail something more, e.g., the person's full emotional life and his or her reactions to the course of the world, including his or her own health status. (These and other aspects of the multidimensionality of the concept of health are discussed in Fagot-Largeault's essay [6].)

Quality of life and human needs

A rather different approach to characterize the concept of quality of life involves the idea of a *human need*. Need has often been thought to be an attractive concept since it suggests an objective state of affairs. It has been favored as the anchoring concept in many social-work and health-care discussions, in particular concerning the allocation of resources. For instance, the Swedish Public Health Act maintains that the policies of the health care system should be determined only by the needs of patients.

There is a long tradition in psychology (following Abraham Maslow [12]) of attempts to characterize the basic needs of a human being. The presupposition is that there is a restricted set of universal, basic human needs. These needs can be identified and studied, it is assumed, by empirical biology and psychology, since these needs continuously manifest themselves in overt behavior.

Anton Aggernaes [1] works quite clearly within the Maslowian framework. In his essay he sets himself the task of identifying a limited set of universal human needs. As a result, he proposes a list of needs which is slightly different from and simpler than the classical Maslowian one. Aggernaes' list of basic needs is exhausted by the following four categories:

1. the elementary biological needs;
2. a need for warm interaction with other human beings;

3.   a need to engage in meaningful activities;
4.   a need for a varied and to some extent exciting and interesting life.

Aggernaes uses the notion of need creatively for the purpose of clarifying the idea of health-related QoL. What Aggernaes calls "objective QoL" is in fact identified with the degree to which the individual has realized his or her fundamental needs. Aggernaes also claims that the individual's "subjective QoL", i.e., his or her satisfaction with life, is analytically linked to the realization of needs.

Moum [14] discusses the concept of need in the context of QoL-research in the social sciences. His main objective is to show that there is no inherent conflict between a "subjectivist" or utilitarian approach to human welfare and a conception that emphasizes needs, rights, and resources. He thus proposes a definition of 'need' based on the notion of happiness. His definition runs as follows:

If it can be shown that a person P obtains more happiness or utility from a given state of affairs A than from another possible state of affairs B, then P may be said to have a need for A (relative to B) ([14], p. 82).

This rather minimal definition of need is supplemented with a number of rules or guidelines for the selection of particular needs as especially worthy of attention, by, e.g., policy makers.

A more theoretical and critical approach is taken by Liss [11], who emphasizes the logical structure of the concept of need. The concept of need is at bottom a relational four-place notion. When agent A has a need, then A has a need for something (the object of need) in order to attain a state of affairs (the goal of need), given certain circumstances. The important impact of this analysis is the following: Unless all the terms in this relation are specified, then the use of the term 'need' is, strictly speaking, meaningless. If one does not know what a particular object — say food, love, or self-realization — is necessary for, then it remains pointless to claim that there is a need for people to have food, love, or self-realization.

Working from this theoretical basis Liss here tackles the notion of QoL. He discusses the idea that the notion of need and QoL could be analytically related. There are then two main possibilities. One is that need should be defined in terms of QoL in the following way: A person can be said to have a need for x if and only if, x is necessary for this person's having a minimal degree of QoL. The other interpretation is the following: A person has a minimal (or even a high) degree of QoL if and only if, all the needs of the person have been realized.

SUMMARY

In this Introduction I have presented the most basic (and in a sense the most important) issues in QoL research, as well as in practical assessment of QoL. The fundamental questions to be asked by a person who intends to engage in these enterprises are the following:

1. What is the purpose of assessing or measuring QoL? Is the purpose, for instance, to obtain a platform for global political decisions? Or, is the purpose to evaluate a medical technology for the rational allocation of funds in health care? Or, is the purpose to find the most adequate treatment for a particular patient in the clinical context?

2. Given this purpose, what aspects of a person's life should be considered? Should we consider a multitude of aspects, containing both personal and circumstantial facts? Or, should we concentrate on selected particular factors?

3. Given this purpose, what evaluative dimension is relevant? A specification of this question is: Should the subject him- or herself decide which is the relevant evaluative dimension?

4. Assume that the aspects of life and the evaluative dimension are provided, what should be the procedure to determine the particular scale for assessing and measuring QoL? Should the procedure be paternalistic? Should it then be *a priori*, based on a particular theory of QoL, or should it be democratic, based on an empirical investigation of people's preferences? Or, should the procedure be individualistic, i.e., exclusively pay attention to the individual's own life-evaluation according to his or her personal normative system?

There seem to be sound arguments for concluding that every serious QoL-researcher should address all of these questions.

BIBLIOGRAPHY

[1]   Aggernaes, A.: 1994, 'On General and Need-Related Quality of Life', in this volume, pp. 241-255.

[2]   Björk, S. and Roos, P.: 1994, 'An Alternative Method of Analyzing Changes in Health-Related Quality of Life', in this volume, pp. 229-240.

[3]   Boorse, C.: 1977, 'Health as a Theoretical Concept', *Philosophy of Science*, **44**, 542-573.

[4]   Bury, M.: 1994,' Quality of Life — Why Now? A Sociological View', in this volume, pp. 117-134.

[5]   Cattorini, P. and Mordacci, R.: 1994, 'Happiness, Life and Quality of Life: A Commentary on Nordenfelt's Towards a Theory of Happiness', in this volume, pp. 59-62.

[6]   Fagot-Largeault, A.: 1994, 'Reflections on the Notion of Quality of Life', in this volume, pp. 135-160.

[7]   Fitzpatrick, R. and Albrecht, G.: 1994, 'The Plausibility of Quality of Life Measures in Different Domains of Health Care', in this volume, pp. 201-227.

[8]   Hunt, S.M. and McEwen, J.: 1980, 'The Development of a Subjective Health Indicator', *Sociology of Health and Illness*, **2**, 231-246.

[9]   Kajandi, M., Brattlöf, L. and Söderlind, A.: 1983, *Livskvalitet: Beräkningar av ett livskvalitets-instruments reliabilitet*, Dept. of Psychology, The Research Clinic, Ulleråker University Hospital, Uppsala, Sweden.

[10]   Kajandi, M.: 1994, 'A Psychiatric and Interactional Perspective on the Quality of Life', in this volume, pp. 257-276.

[11]   Liss, P-E.: 1994, 'On Need and Quality of Life', in this volume, pp. 63-78.

[12]   Maslow, A.: 1954, *Motivation and Personality*, Harper & Row, New York.

[13]   McDowell, I. and Newell, C.: 1987, *Measuring Health: A Guide to Rating Scales and Questionnaires*, Oxford University Press, Oxford.

[14]   Moum, T.: 1994, 'The Role of Needs, Rights and Resources in Quality of Life Research', in this volume, pp. 79-93.

[15]   Musschenga, A.W.: 1994, 'Quality of Life and Handicapped People', in this volume, pp. 181-198.

[16]   Naess, S.: 1987, *Quality of Life Research: Concepts, Methods and Applications*, Institute of Applied Social Research, Oslo, Norway.

[17]   Naess, S.: 1994, 'Does Self-Deception Enhance the Quality of Life?', in this volume, pp. 95-114.

[18]   Nordenfelt, L.: 1993, *Quality of Life, Health and Happiness*, Avebury, Aldershot, U.K.

[19]   Nordenfelt, L.: 1994, 'Towards a Theory of Happiness: A Subjectivist Notion of Quality of Life', in this volume, pp. 35-57.

[20]   Nozick, R.: 1989, *The Examined Life*, Simon and Schuster, New York.

[21]   Nussbaum, M.: 1993, 'Non-relative Virtues: An Aristotelian Approach', in Nussbaum, M. and Sen, A.(eds.), *Quality of Life*, Clarendon Press, Oxford.

[22]   Ostenfeld, E.: 1994, 'Aristotle on the Good Life and Quality of Life', in this volume, pp. 19-34.

[23]   Rosser, R. and Kind, P.: 1978, 'A Scale of Valuations of States of Illness: Is There a Social Consensus?', *International Journal of Epidemiology*, 7, No 4, 347-358.

[24]   Sandøe, P. and Kappel, K.: 1994, 'The Plausibility of Quality of Life Measures in Different Domains of Health Care', in this volume, pp. 161-180.

[25]   von Wright, G.H.: 1963, *The Varieties of Goodness*, Routledge & Kegan Paul, London.

SECTION I

THE CONCEPT OF QUALITY OF LIFE

ERIK OSTENFELD

ARISTOTLE ON THE GOOD LIFE AND QUALITY OF LIFE

Aristotle thinks[1] that there is some good at which all actions aim. There must be some end which is wanted for its own sake, and for the sake of which we want all the other ends (if we do not choose everything for the sake of something else — which involves an infinite regress). This is the *summum bonum* which it is important to know in order to plan our lives. The science that studies that end is politics (political science)[2]. This is the the most authoritative and directive science. It makes use of the other sciences and its end, the human good, includes their ends. The good of the community is greater and more perfect than that of the individual, and man is a political animal and needs social community with others, i.e., his individual good is imperfect in itself.

In Aristotle's view the good life is the most desirable life[3]. Hence an exploration of the good life, i.e., the best (not necessarily morally best) or most worthwhile, desirable life involves an examination of the following concepts and their interrelation: the good, happiness, virtue, pleasure, welfare, society and political constitution. What are the necessary and sufficient conditions for applying these concepts? How do they relate?

In view of this frame of reference, it seems clear that if we want to know how Aristotle would define quality of life (either in the sense of emotional response or of experienced preference satisfaction), we must look at his conception of the good life. This will include attention to how the notion of pleasure (a most notable experience *EN* 1099a8,1170b1) is related to the other concepts mentioned, in particular "happiness" and "welfare".

We cannot expect an accurate account of the principles of politics (admirable and just actions and goods), since they are so only for the most part. We must rely on the unclear opinions (cf *EE* I,6) of the many and the wise who agree in saying that the end is *eudaimonia*[4]. By this they mean living well or doing well (*EN* 1095a18-20, cf *MM* 1184b). They are not agreed, however, whether this involves bodily pleasure, money, honor or something else. A wise man like Plato held that the highest good was the idea of the good.

But bodily pleasure is the end of animals, honor is dependent on those who give it and not proper to its possessor and is often a means to appearing

L. Nordenfelt (ed.), Concepts and Measurement of Quality of Life in Health Care, 19–34.
© 1994 Kluwer Academic Publishers. Printed in the Netherlands.

good. Goodness in that case is compatible with inactivity and misfortune. Wealth again is a means to something else.

As for the idea of good, there is no common idea of good in the categories: there would have to be one single science of all good things and the notion of a transcendent idea is unintelligible. Anyway it is useless in practice, it cannot be realized and is not used even as a pattern (cf *EE* I 8, *MM* 1182b10-1183b19).

For Aristotle, if there is any one thing that is the end of all actions, this will be the practical good - or goods, if there are more than one. However, as there are more ends and we choose some as means to others the best for man is thought to be (1) something *complete* and *final* (*teleion*)[5], and (2) something *self-sufficient* (*autarkes*) (cf *EN* I 7)[6].

Now x is "more final" than y if it is pursued for its own sake (while y is pursued because of something else), or it is never choosable because of another (while y is choosable because of x as well as for its own sake), and it is final without qualification if it is always choosable for its own sake and never because of something else.

Eudaimonia seems to be an end that is always chosen for itself and never because of something else, whereas honor, pleasure, intelligence and the virtues are chosen partly for themselves partly for their consequence, i.e., eudaimonia. Hence eudaimonia fulfils one of the marks of the highest good.

The second mark is self-sufficiency, which alone makes life choosable and lacking in nothing (*EN* 1097b14-5). As man is a social being a self-sufficient life would involve parents, wife and children, friends and fellow-citizens. Eudaimonia is such a self-sufficient thing too. It is the most desirable (choosable) of all things, but not as one of many goods (as this would make it yet more desirable if one of these were added). The *MM* 1184a19 emphasizes the composite nature of eudaimonia.

Aristotle fills in this conceptual scheme with a substantial notion of eudaimonia. He suggests that we look at the function of man. Just as the good of the craftsmen lies in their function so it may be the case that the human good in general can be found in the function of man, if indeed such there is. He advances two arguments to prove that man has a function: 1) the analogy of craftsmen, 2) the analogy of bodily organs and parts like eye, hand, foot, etc. He excludes life which is shared by plants, i.e., the nutritive life and growth, and also sensitive life shared by other animals. But a practical active life of the rational part is special for man. Hence man's *function* is said to be an *activity of the soul in accordance with reason* and the human *good* an *activity of soul in accordance with virtue* (=excellence), and if there are more, *in accordance with the best and most complete virtue* (*EN* 1098a3-20)[7].

The function argument which argues that just as a good lyre-player performs well on the lyre, so a good man performs well as man (*EN* 1097b24-98a20) has been attacked as a false analogy: craftsmen have tasks by definition, but man does not have a function at all or at least not a clear, uncontroversial one. Even if reason is unique to man it is not thereby the best, and certainly not all expressions of human reason are valuable. And even if he had a clear task, being good at being human would not make him a good man (cf. the *techne* analogy[8] and the ambiguity of "living well and doing well"[9]). The passage should be read in the light of *EE* 1218b37-1219a39: the virtue of anything is defined as the best state, condition or faculty of all things that have a use and work. Thus garments, ships and houses have virtues or excellences[10], and so has the soul, for it has a work, i.e., to make alive. And the work of anything is its end, that for which it exists. The work is either separate from (e.g., a house or health) or identical with the use (*chresis*) (e.g., seeing or contemplation).

The definition of happiness is shown to be in agreement with common views on happiness. First, it is a mental good (an activity or action belonging to the soul). Second, it implies living and faring well (cf. the common analogy with health[11]). Third, it includes or involves the typical candidates like virtue, prudence, wisdom (cf *EE* I 5), pleasure and external goods. Obviously, the virtues are included, being directly mentioned in the definition (though we could question whether Aristotle's definition includes virtue in the moral sense), but also pleasure is a part of the definition of eudaimonia in so far as the virtuous person finds pleasure in virtuous actions (obviously it is purer pleasures that go with fine actions[12]). These are not only good and fine but also pleasant in themselves. Hence happiness is the best, the finest and the most pleasant of all things. It does not need to have (bodily or otherwise) pleasure attached: fine actions are naturally pleasant. Only a certain amount of external goods or good fortune (*eutychia*) seems required. Fine actions in themselves presuppose to some extent wealth, friends and influence, and felicity presupposes a good family and at least lack of ugliness if not actual beauty.

Aristotle conceives of virtuous activity as the essence of happiness, and all other goods as either necessary preconditions or natural contributors/instruments (*EN* 1099b25-7). We may assume that good fortune would count as necessary precondition, while pleasure in good action would go as contributor[13]. The virtuous person is to some extent above the vicissitudes of fortune because of the permanence inherent in virtuous activity. He will spend all his life in virtuous conduct and contemplation and bear what may befall him. Within reasonable limits. No one could be happy on the rack (1153b19-21). The truly happy, however, will never become miserable as he will never do things that are hateful and mean. But it must be admitted that grave misfortune would prevent the virtuous from being entirely happy. Hence Aristotle gives as his final definition of happiness: activity in accordance with complete virtue, adequately furnished with external goods, throughout a complete life (*EN* 1101a14-16).

As we have seen, eudaimonia for Aristotle is living well and doing well (*euzoia* and *eupraxia*). This amounts to success and self-realization. The *Ethics* is a manual on how to get the best, finest and pleasantest life (*EN* 1095a5-6, 1103b26-29, cf. *EE* 1216b). The "best" refers to the objective realization of man's potentialities, the "finest" refers to the respect shown by one's fellow men (the moral aspect), and the "pleasantest" refers (partly) to the subjective satisfaction to be got from this life (perhaps what is meant by "happiness" in the contemporary debate). Eudaimonia is then a much richer concept than the modern concept of *happiness* in so far as that is understood as a feeling or emotion. It is not just a subjective experience, but also a realization of the person's mental potential, feelings as well as intellect. The former are disciplined into a mean state by reason (*EN* II 1 and 6), while the intellect is taught (*EN* 1103a14-7) to reach a contemplative state and to guide man's action. Finally it is the admired (moral) ideal of one's fellow citizens, what has been called the normative conception in contrast to the conative conception of the final good (Irwin 363).

We shall now take a look at the most subjective ingredient in happiness, pleasure. Aristotle considers that it is an experience. However, as we shall see, it is not simply a subjective phenomenon. It is basically what Aristotle call an activity.

Pleasure is involved at different levels in the *Ethics*: virtue and political science in general are concerned with pleasure/pain (*EN* II 3 and 6, 1152b1-8), self-control with pleasures of touch and taste (*EN* III 10-12), and connected with this is the pleasure of incontinence (*EN* VII 4 and 7). Friendship may be grounded in pleasure. Finally, good acts are done with pleasure and there is the pleasure of the gods and, in happy hours, of men (contemplation).

Aristotle's view(s?) of the nature of pleasure appear(s) to some extent from his treatment of the above-mentioned problems, but especially from two longer passages devoted especially to this topic: *EN* VII 11-14 and X 1-

5. The former consists mainly of a refutation of the following views: no pleasure is good, most pleasures are bad, and pleasure is not the highest good. During this refutation we learn that pleasure is not a process but a kind of activity and end, or more precisely that it is not a felt process, but *unimpeded activity of a natural state*. Pleasure might even be the highest good in so far as it is the unimpeded activity of our capacities and in view of the fact that all animals pursue it.

In *EN* X 1-5 we find Aristotle's maturest, most balanced and positive (non-dialectical) view. He first opposes both the idea that pleasure is the good and that it is not a good. While refuting the last view he argues that pleasure is not a movement, because every movement has its own speed and one cannot be pleased quickly (though one may become pleased quickly). Nor is it a process, e.g., a replenishment of the (empty) body. There are pleasures of learning and of smell and certain sights and sounds, memories and hopes that do not have antecedent pains, and it is not clear of what they could be the coming into being.

Aristotle's positive view is found in chs 4-5. It is claimed that pleasure is whole, perfect at any moment, like the act of seeing. Hence it is not a process, which is in time and has an end (as the process of building) and is complete only in the whole of time or at the instant of reaching its end (walking is another example). The form of pleasure is complete at any given moment. Movement occupies time, whereas pleasure does not. For that which "takes place in a moment is a whole." "Movement" and "process" apply to things consisting of parts. But seeing, a point, a unit and pleasure neither result from a process nor are they processes, but they are wholes (cf *Met* 1048b18-35). Now what Aristotle seems to be saying is that pleasure is like seeing in being outside time in the sense that it does not occupy time[14], it need not cease, speed does not apply; in short, it is not a physical process or a movement. Aristotle has an idea of divine "activity of immobility" (*EN* 1154b27) and he thinks that (true) pleasure is "more in rest than in move- ment", and this it shares with seeing, thinking, living well or being happy and living (which is understood as primarily sensing and thinking (*EN* 1170a19)). We seem to have here a non-reductionist view of man, who is seen as more than sheer matter in motion.

Moreover, the activity of any sense is at its best when the organ is in the best condition and directed towards the best of the objects proper to that sense, and similarly for thought (except that thought has no physical organ). This activity will be most perfect and most pleasurable; for there is a pleasure corresponding to each of the senses, just as there is to thought and contemplation. Pleasure perfects the activity, but not in the same way that good organs and objects do. It is a sort of *supervening perfection*, like the bloom that graces the flower of youth.

The reason why everyone seeks pleasure is that everyone wants to live (an activity) and pleasure perfects this activity. Hence we may say that we choose life because of pleasure or pleasure because of life. It does not matter as they are connected and do not admit of separation. Pleasure does not occur without activity, and perfect activity does not occur without pleasure. As we saw above, pleasure is not a feeling or sensation resulting from or caused by an activity. But it is also absurd to think that pleasure is identical with the activity (*EN* 1175b33-6). What then is its status?

Pleasure was defined in *EN* book 7 as (a) *the unimpeded activity of a natural state* (1153a14-15). However, in book X we learn that being conscious that one is alive is pleasant in itself because life is by nature good and to be conscious that one possesses a good thing is pleasant (1170b1-5). Life is itself a form of activity, and pleasure is now said to perfect the activity of life (1175a10 ff). Pleasure does not perfect the activity as the formed state that issues in that activity perfects it, by being immanent in it (cf. e.g., the natural function of the sense-organs), but as (b) a sort of *supervening perfection*, like the bloom that graces the flower of youth (1174b31-33). Pleasure and activity (here life) are nevertheless closely connected and do not admit of separation, they occur together (1175a19-21, 1175b32-36). There seems to be an inconsistency in Aristotle's conception of pleasure: is it (1153a14-5) or is it not an activity (1175b34-5)? (A) says that pleasure is unhindered activity, while (b) begins by likening pleasure to seeing (activity) (1174a13-b14), but ends by claiming that it perfects activity and that it is *not* identical with that activity. How can these claims be reconciled or explained? It has been suggested that Aristotle refines his account (perhaps, but not necessarily, a historical development). Alternatively, it has been assumed that in book 7 Aristotle deals with the objective base of pleasure (the formed state issuing in activity, immanent in the thing) while in book 10 he turns to the concept of pleasure[15].

At this juncture perhaps the following observations should be made: pleasure seems to play some causal role (*teleioi* 1174b23,31, 1175a17,21, *synauxei* 30-1), but at the same time it "follows" upon activity (1174a7, 1175a6). The clue may lie in the role of consciousness (1170b1-10). The awareness that one possesses a good (e.g., life) is pleasant in itself. This pleasant awareness itself improves the activity as a final cause and pleasure may be said to "follow up" that activity, when one realizes that one acts successfully. The improved activity in turn becomes the object of pleasant awareneass and is thus further improved, etc. The *final* cause is not, in Aristotle's view, different from the *formal* cause, nor is it identical with it (to begin with). Hence, if this is at all a plausible interpretation, refinement is perhaps what characterizes the treatment of *EN* X in relation to *EN* VII.

However, "pleasure" is *ambiguous* (and so is the Greek term *hedone*) between the sensation of pleasure and the source of pleasure (1176a18-9). Moreover, sensations of pleasures *differ* as the activities to which they are attached differ: as Aristotle puts it, pleasures differ in kind and source[16] (1174a10-11). Pleasure can only be identified by reference to the activity it promotes. Things that are different in kind are perfected by things that are different in kind, and what perfects one kind of activity must differ in kind from what perfects another. Now the activities of the intellect differ in kind from those of the senses, and both differ among themselves; therefore the pleasures that perfect them also differ. The same may be seen from the fact that the pleasure proper to an activity intensifies it. If one enjoys an activity one becomes good at it, while alien pleasures are destructive.

According to *EN* book 10, pleasures differ as they come from (*apo*) noble acts or base acts (1073b23-9). Activities differ in goodness and badness, and some are to be chosen, some to be avoided, some are neutral. Therefore their proper pleasures can be classed similarly. Thus pleasure proper to serious activity is virtuous, whereas pleasure proper to bad activity is vicious.

Every animal species has a proper function and a corresponding proper pleasure of exercising that function. Hence men do not take pleasure in the same things that dogs do, but dogs among themselves presumably find the same things pleasant. Human beings, however, do differ in their preferences. Here the good man's view is the true one. The *true* pleasures then become those that *seem* to the good man to be pleasures and those things will be really pleasant that he enjoys. The disreputable pleasures are not pleasures at all, and of the reputable pleasures only those are human that *perfect* the one or more activities of the perfect man. The rest are human in a secondary degree (*EN* X 5).

As sight is superior to touch, and hearing and smell to taste in purity, so their pleasures differ similarly. Intellectual pleasures are superior to sensuous one.

Apart from the distinctions mentioned above according to moral value, specific function and purity Aristotle distinguishes between the following sources of pleasure: *bodily* (love of sensing, especially of touch) and *mental* pleasures where the body is not directly affected (e.g., love of money, honour, gossip or thinking) (1117b28-99). A related distinction is that between *necessary* or *physical* pleasures (needs such as food and sex) and pleasures *desirable in themselves* (victory, honor, wealth and other good things 1147b25-31, cf 1176b2-3). The latter are also called *naturally desirable* (1147a24,1148b3) or naturally pleasant. Of the latter some are pleasant absolutely (fine actions *EN* 1199a13-14), whereas others are pleasant only to certain kinds of men and animals (1148b15-16). Pleasures unaccompanied by pain are derived from what is pleasant by nature (those that stimulate the activity of a natural disposition, e.g., music or contemplation), whereas remedial processes only seem pleasant, i.e., they are only *incidentally pleasant*, the really pleasant being the activity of the part of the organism that has remained healthy (1154b16 ff, cf 1152b33 ff).

How then can pleasures be compared and ranked as more or less pleasurable? How can we tell that contemplation is the most pleasurable activity? We cannot say that contemplation gives more of one stuff, pleasure, than, say, eating. A hedonistic calculus cannot be established as pleasures differ in kind. And yet nobody would choose to live out his life with the mentality of a child, even if he continued to take the greatest pleasure in the things that children like. Nor would one choose to find enjoyment in doing something very disgraceful, even if there were no prospect of painful consequences (1174a1-5). What is the criterion of choice here?

In fact two questions may arise here: How do we compare and rank different pleasures? And secondly: Who is to judge? It seems significant that Aristotle answers both the first (objective) question and the second (subjective) one (cf. below p. 29). One possible answer to the first question is: completeness (*teleiotes*). The most *perfect* activity is most pleasurable. But complete or perfect sense-activity obtains when the organ is in the best condition and directed towards the best of the objects proper to that sense. This can be generalized to cover every human activity that is exercised upon an object, e.g., thought and contemplation (1174b14-1175a3). That the faculty must be in a good condition seems straightforward enough, but what is the exact import of the requirement that the object be best? Strictly, it should be most beautiful (*kalliston* 1174b15), best (*kratiston* b19), most worthy (*spoudaiotaton* b22-3) and as it should be (b34). What could that be? The best of its kind?[17]

It may be suggested, however, that one activity, e.g., sight is more complete than another, e.g., touch. This would (in cognitive activities) be connected with the nature of their proper object. What, in that case, would count as a most perfect object? A *simple,* invariable (i.e., immaterial) object perhaps? Certainly, if we were simple ourselves the same activity would always give us (as God) the greatest simple pleasure. For there is an activity not only of movement, but also one of immobility; and there is a truer pleasure in rest than in motion (1154b20-8). The concern here is the subject rather than the object, but the reference to God's activity makes the inference from activity to object valid: God has his own activity for object, divine thought thinks of itself, since it is the most excellent of things (*kratiston Met.* 1074b34, *ariston Met.* 1072b14-19). His contemplative activity is the most pleasant (*ibid.* 24). Now if this is what a most perfect object is, it remains to be seen how the objects of sight are more simple and invariable than the objects of touch. As a first approximation it may be noted that sight is more discriminatory than touch. It makes a wider range of distinctions than touch. Another way of putting this would be to say that sight is more selective and its objects more abstract and therefore more simple and invariable.

There is a related and probably basically identical criterion that classes activities according to *purity*: sight is thus purer than touch, hearing and smell than taste, and intellectual activity than sensuous activity. And their pleasures differ correspondingly (1175b36 ff). Thus physical pleasures are distinguished from pure and refined pleasures connected with virtue and intelligence (1176b18-21). Physical pleasures are food, drink and sex, whereas pure and refined pleasures are those attending virtue and contemplation. How are we to understand purity on the part of object if not in terms of relative simplicity and unchangeability (as above)?

Turning now from the object to the activity: Pleasures, as we have seen, may be of two kinds, pure and mixed (1173a23-25), the former being determinate, the latter indeterminate. The determinate is good and pleasant (1170a19-20) and pleasure, Aristotle suggests, could be *determinate* just as health[18], although it admits of degrees.

Moreover, an activity is pleasurable in so far as it is *continuous* (1170a7) and *unimpeded* (1153b11). Here contemplation (*theoria*) is unrivalled. One part of our intellect (the scientific) is capable of contemplating those things whose first principles are invariable, another (the calculative) contemplates things that are variable (1139a6-12). Now in the final pages of his *Ethics* (*EN* X 7-8) Aristotle turns back to his early statement that happiness is the activity of the soul according to virtue, and if there are more than one, according to the best and most complete (1098a16-18). The best virtue must belong to the best part of us, which is a naturally ruling part and one that has insight in to fine and divine matters. It is either divine or more divine than any other part of us, and it is the contemplative activity of this in accordance with its virtue that will be happiness. Aristotle advances a series of arguments to support this surprising view: 1) the activity is best since the intellect and its object are best, 2) it is most continuous and 3) most pleasant (pure and lasting pleasures), 4) most self-sufficient, because one can contemplate by oneself, 5) it is appreciated for its own sake, nothing else is gained by contemplation, 6) it is the most leisure-like activity. Hence, provided that it is allowed a complete term of life, contemplation is perfect happiness (*EN* 1177b24-5). However, the intellect appears to be something divine, superior to our composite nature, and its life divine in comparison with human life. Still, we should live the contemplative life of the authoritative and best part which is our true self. Life of practical reason according to prudence and the moral virtues is human life and yields human happiness. But the perfect happiness of the intellect is separate (*kechorismene*). Further arguments are: contemplation is the sole activity of the gods, who are supposedly supremely happy and blessed living beings. Also, none of the other animals are capable of being happy, having no capacity for contemplation.

If we ask what are the objects of contemplation then it must be inferred from book VI that they are the things whose principles are invariable. According to *EN* VI 7 wisdom (*sophia*) is knowledge of the most precious truths. It is scientific and intuitive knowledge of what is by nature most precious. Now intuition apprehends first principles, while science judges about the universal and necessary (*EN* VI 6), and if necessary, then eternal (VI 3). It should be noted, however, that contemplation does not include search for truth (1177a26), and this excludes normal scientific work. This is all the *EN* has to tell about the nature of the objects. But it may be supplemented by the *Metaphysics* (1005b1, 1026a18), where we learn that there are three theoretical sciences, first philosophy (theology), mathematics and physics. We may assume that it is only in so far as physical events have an invariable aspect that they may become an object of contemplation. Contemplation may have religious overtones in so far as it covers the objects belonging to theology (astronomy and the universe as a whole)[19].

It may be wondered what the aim of morality then should be? Moral virtue or contemplation? Or both, and in that case how? There can be no doubt that Aristotle has not written his *Ethics* largely about the moral virtues only to discard them in the end. Both the practical (human) life and the theoretical (divine) life must have a place if man shall reach happiness.

However, it has been argued that, since human morality cannot just be done in one's spare-time or as merely instrumental to one's real interest, Aristotle's vision reflects a contemporary aristocratic ideal of an elite of philosophers kept alive thanks to the more human life of a supporting community. Such a sharing of work is not unknown in our own societies. Alternatively, we must conclude that Aristotle just did not reconcile his two lives, owing to doubts about the nature of man, a monistic, biological or a more Platonist, dualistic view.[20]

Looking back at the argument so far, some general comments are in order. First, it is evident that Aristotle does not advocate ethical hedonism. Pleasures differ in kind and may be classed as good or bad. Not all and only pleasure is intrinsically good (e.g., moral qualities are intrinsically good). Hence pleasure and the good are different (*EE* 1235b23). However, absolutely they are identical: the good is necessarily pleasant (the good man is not good if not pleased by the good *EN* 1099a17, cf *EE* 1236a5-7). Fine actions are both desirable in themselves (1144a1, 1176b8-9) and pleasant in themselves or by nature (1099a21, *EE* 1237a6-7). Why else should people want to do them? (1099a7-16, 1154a1-7, 1152b6-7, 1153b29-32, 1172a28?). The absolutely good is identical with the absolutely pleasant (*EE* 1235b2-3, cf 1236a 5-7), and the absolutely pleasant is fine (*kalon EE* 1249a18-19). But we would still choose something even without pleasure (*EN* 1174a4 ff), e.g., sight, memory, knowledge and the several excellences, although as a matter of fact they do produce eudaimonia (1144a1 ff).

Second, there is both an objective and a subjective criterion for identifying the true pleasures. They are the pleasures (1) that *perfect* the activities of the perfect man, i.e., those that accompany man's function, (the pure, determinate and continuous pleasures and (2) that the good man *thinks* are pleasures. They are the pleasures of virtuous action and contemplation, which the prudent man knows are truly pleasant. There is not only overlap but identity between what is objectively and what is subjectively classified as pleasant. Just as the good in nature is good to the good man (1170a21-22), so the naturally pleasant is pleasant to the good man (1113a30-4).

To the second question raised earlier — Who is the criterion of pleasantness? — the answer is: the good man[21]. But is this not a non-Aristotelian subjective criterion after all? Not in so far as Aristotle insists (*EN* 1113a29-33) that what is pleasant (and good (1170a14-6) and fine) to the good man is really pleasant (and good and fine). There is correspondence between phenomena and reality in the case of the good person[22]. The real pleasures and really good things seem here to be defined by reference to the proper function of human nature or to a general theory of health. This is not circular, however, in so far as the proper function is defined in general terms, not again in terms of real pleasure, etc. Also, we found several independent objective criteria for identifying true pleasure above. Still, it is understandable that circularity has been suspected in so far as we find the following definitions: the good is what the good man thinks is good (*EN* 1170a14-16, cf 1113a25 ff) and the good man is the one to whom the naturally good seems good (*EE* 1248b26-7)[23].

Pleasure is not only a subjective phenomenon. We take pleasure in what is *our own* (1169b33, 1177b24, 1178a5-6, cf 1099a8-9), and what is our own is our proper activity. The pleasure we get from that is defined in terms of that activity[24]. So it is not (just) a sensation. Aristotle seems closer to a motivational theory of pleasure which posits a logical connection between pleasure and desire and/or doing. We know off hand what pleases us in a way we do not know what is the cause of, say, a pain in the stomach. Emotion comes of doing and leads necessarily to more doing. Hence emotions are identified by their objects. In so far as pleasure is activity it is an objective phenomenon.[25]

In conclusion: everybody desires happiness or welfare (eudaimonia). But what is eudaimonia? A disposition, or state, or sensation, or emotion, or activity? Answer: an optimal mental activity according to our special human capabilities. A doing rather than an undergoing (or state of mind). Complete in itself and a composite good (*EE* 1184a), most admired, useful, and pleasant of all (*EE* 1214a). More than the sum of immediate satisfactions, but also more than long-term satisfaction in so far as that satisfaction is only understandable as being "with" an object. Eudaimonia is rather the working and self-realization of the human being, which is the good for man, the most satisfactory, desirable (*hairetotaton*) highest good (*summum bonum*) or value.

The foundation of good (and morality and life-quality) is function (nature). The good is defined in terms of objective human capabilities and the conditions for their maximal exercise. Coupled with this is pleasure, which is objective in so far as it is identified (not exactly with but) *via* succesful activity. That it also has a subjective component is obvious, but that is not the point of departure for Aristotle.

By saying this it is obvious that Aristotle is not in the post-Cartesian trend emphasizing experience as the arbiter of everything, including quality of life. He is definitely, and perhaps paradoxically in view of the common picture of him, not even an empiricist (not to say a utilitarian) in his philosophy of eudaimonia. On the contrary, building on a metaphysical view of man as a bundle of physical and mental capabilities he claims that eudaimonia (human flourishing) is the realization of mental capabilities (though certain external goods, a good fortune, are required too). What then does the realization of mental capacities, i.e., activity in accordance with complete virtue[26] mean? It means a realization of both feelings and intellect, both practical and theoretical (though it must be admitted that the theoretical intellect is focused on in *The Nicomachean Ethics* X). The flourishing man or woman to Aristotle is a person who is well-adapted and well-functioning socially and psychologically, and who lives a rich life in the sense that he or she uses all human potentialities, i.e., not restricted to immediate pleasure seeking from e.g., television watching, beer-drinking and easy sex, and not even a more calculated pleasure seeking, but involving intellect and feelings that are special to man. This is to say, a life in contemplation of truth and as a responsible citizen and part of a society and a family. Such a life is claimed to be extremely pleasant because it is pleasant for man to exercise his or her abilities. Moreover, we recognize such a person as 'admirable', as Aristotle says, because he or she knows how to live, how to have success as a human being. Accordingly, eudaimonia is measurable, in so far as realization of specifically human capabilities is objective.

It should be obvious by now that Aristotle may be relevant to the modern debate on quality of life, which is labouring in the Cartesian shadow under the incompatible requirements of counting and measuring everything and at the same time wanting to take the first person (subjective) point of view (happiness= feeling) . This leads of course to over-democratic definitions of quality of life (emotional response to life-situation or experienced preference satisfaction or the like) whence come the well-known problems of measuring and comparing quality of life. The social and medical engineers naturally want to know whether they have been successful, but they cannot be given the guidelines they need because no objective definition of quality of life can be constructed on the basis of the prevailing subjectivist ideology. What Aristotle tells us is that satisfaction and well-being are is to be found within a frame of reference that is biologically and sociologically defined. He offers us 'eudaimonia' rather than 'happiness', i.e. quite another conceptual framework. Perhaps this is what is needed to soften up present day's conversation on quality of life? We surely need to steer equally clear of the barren scepticism of extreme subjectivism and of the mad engineer's vision of paradise.

NOTES

[1] I shall base my account mainly on Aristotle's latest and most mature course in ethics, *The Nicomachean Ethics* (*EN*), available in *Aristotle's Ethics*, translated by J.A.K.Thomson, rev. ed. by J. Barnes [3]. The Greek text used here is I. Bywater's ed. in Oxford Classical Texts, London 1894.

[2] Statesmanship is an art (*techne*) that aims at making men good (Plato, *Protagoras, Gorgias* 515c,517c, *Republic* 420b, *Politicus* 305e and Aristotle, *EN* 1103b2-5).

[3] Cf *Eudemian Ethics* (*EE*) 1215b17-18 (the *EE* is now generally agreed to be by Aristotle and to be an earlier version of the course on ethics that found its final expression in the *Nicomachean Ethics*) and *EN* 1097a25-b21, 1176b4-6 (the good life or happiness is chosen for itself). Thus far Aristotle agrees with Eudoxus: that which is good for all and which all try to obtain is the Good (*EN* 1172b14-5), but he cannot accept that: in every case what is desirable is good (*ibid* 10-1), which is reminiscent of J.S. Mill.

[4] human flourishing or welfare (*euzoia*), success (*eupraxia*), living blissfully and admirably (*EE* 1214a30), or living admirably (*EE* 1214b8,16), or blissful and happy (*EN* 1098a19). Eudaimonia is at once the most admirable, the best and the pleasantest (*EE* 1214a7-8, *EN* 1099a27-28). For Socrates it could be expressed by "having a good fortune" (*eutychein* in Plato, *Euthyd*. 278e-282d).

[5] See [11], pp. 361, 605 , [1], p. 23f, and [7], 384 f.

[6] In *Magna Moralia* (*MM*) 1184a a complete end is defined as one whose attainment wholly satisfies us, so that we do not long for something else. It is the fulfilment of the good. Here condition (1) seems explicated in terms of condition (2) so that we have one not two marks of the best. *MM* is of disputed authorship. There is a tendency now to see it as the earliest Aristotelian sketch of morality.

[7] Cf *EE* 1219b: Eudaimonia is the activity of a complete life according to complete (= the whole of) virtue. It should last for a life-time (being a whole).

[8] The *techne* analogy was used extensively by Socrates in his discussions with sophists on virtue (Plato, e.g. *Rep*.I). The shortcomings of the analogy are discussed in Plato, *Hip. Min*., which was known to Aristotle (*Met* 1025a6, cf also *EN* 1140b21-30,and *EN* II chs.1 and 4 and 1106b8-16).

[9] "Well" may be taken in a moral and a non-moral sense.

[10] Cf *EN* 1106a15 ff: eye and horse have functions.

[11] E.g.*EE* 1216b20-5, 1227b, 1249a21 ff; *EN* 1102a17 ff, 1105b12 ff, 1104a4-5, 1112b, 1114a, VI 12; *MM* 1199b28 ff.

[12] See below and *EE* 1216a26 ff.

[13] See below. *EE* I 2 deals with the problems of 1) distinguishing special necessary conditions as meat eating or post dinner walks from general conditions like breathing, being awake or

moving , 2) distinguishing necessary conditions in general from the essence of happiness. See also Ackrill [1], pp. 245, 267. The problem can also be stated as the conceptual one of specifying the necessary and sufficient conditions for the application of "happiness" (cf *EE* 1214b1-6, *EN* 1098b22,26).

14 Cf. [12], p. 176: not taking time but lasting for a time. For this nuance see *EE* 1238a25-6.

15 Cf Owen, "Aristotelian Pleasures" pp. 336, 346, criticized by Gosling and Taylor [9], pp. 204-24.

16 There is a problem here: does Aristotle mean to distinguish pleasures by their sources and/or objects? Kenny [12] discusses how Descartes confused the two (p. 10 f), and how the British empiricists failed to see the intensional nature of the object of emotion (pp. 22 ff, 60 ff). It does not seem that Aristotle does any better. However, the distinction between object and cause is essential because emotions are specified by their objects (pp. 60, 66, 72 f).

17 Cf. [12] p. 148. Gonzales's recent analysis in terms of "beautiful object" is interesting but too speculative to carry conviction.

18 Contra Plato *Philebus*.

19 *Met*. XII, 6-10 and esp *EE* 1249b20. Cf. [15] 232-4.

20 Cf [1] for both interpretations.

21 Cf *EN* 1176a16, 1099a7 ff, 1113a25 ff, 1166a12, 1170a14 ff, 1176b25-6.

22 Cf EE 1235b31-1236a7. [9] pp. 330-344, esp 336 is critical about the possibility of identification of the good man (*phronimos*) independently of pleasant experiences and of a preference free test of the naturally pleasant.

23 A way out beyond the judgement of the good man is suggested by *EE* 1249b17-21: the best and finest standard is whatever choice and possession of things good by nature will best promote the contemplation of God.

24 Cf *EN* 1154b20: things pleasant by nature are those that bring about the activity of such and such a nature, and *EE* 1238a25-26: the absolutely pleasant is defined by the end it effects and the time it lasts.

25 Cf. [12] p. 127.

26 A. Kenny [13] has recently argued that the holistic character of Aristotle's mature ethics is set out in the *Eudemian Ethics*.

BIBLIOGRAPHY

[1]   Ackrill, J.: 1973, *Aristotle's Ethics*, Faber and Faber, London.

[2]   Ackrill, J.: 1974, 'Aristotle on Eudaimonia', in Rorty, A.O. (ed.), *Essays on Aristotle's Ethics*, UCP, 1980.

[3]   Aristotle: 1976, *The Nicomachean Ethics*, transl. by Thomson, J.A.K., rev. by Tredennick, H., introd. and bibl. by Barnes, J., Penguin Books, London.

[4]   Aristotle: 1952, *The Eudemian Ethics*, transl. by Rackham, H., Harvard University Press, Loeb Classical Library, Cambridge, Massachusetts.

[5]   Aristotle: 1935, *The Magna Moralia*, transl. by Armstrong, G.C., Harvard University Press, Loeb Classical Library, Cambridge, Massachusetts.

[6]   Aristotle: 1935, *The Metaphysics*, transl. by Tredennick, H., Harvard University Press, Loeb Classical Library, Cambridge, Massachusetts.

[7]   Cooper, J.M.: 1975, *Reason and the Human Good in Aristotle*, Harvard University Press, Cambridge, Massachusetts.

[8]   Gonzales, F.J.: 1991, 'Aristotle on Pleasure and Perfection', *Phronesis*, vol. XXXVI.

[9]   Gosling, J.B.C. and Taylor, C.C.W.: 1982, *The Greeks on Pleasure*, Clarendon Press, Oxford.

[10]   Hardie, W.F.R.: 1968, *Aristotle's Ethical Theory*, Clarendon Press, Oxford.

[11]   Irwin, T.H.: 1988, *Aristotle's First Principles*, Clarendon Press, Oxford.

[12]   Kenny, A.: 1963, *Action, Emotion and Will*, Routledge and Kegan Paul, London.

[13]   Kenny, A.: 1992, *Aristotle on the Perfect life*, Clarendon Press, Oxford.

[14]   Owen, G.E.L.: 1972, 'Aristotelian Pleasures', in Nussbaum, M. (ed.), *Logic, Science and Dialectic*, Duckworth, London 1986

[15]   Ross, W.D.: 1949, 5. ed., *Aristotle*, Methuen UP, London.

TOWARDS A THEORY OF HAPPINESS: A SUBJECTIVIST NOTION OF QUALITY OF LIFE

INTRODUCTION

My task in this essay is to suggest a characterization of a concept of quality of life which could, hopefully, serve as a conceptual basis for the construction and evaluation of instruments designed for the measurement of quality of life. The concept to be developed is a subjectivist concept, in fact identified with *happiness-with-life*. Happiness is looked upon as a species of — by far the most important species of — *well-being*. Well-being is in its turn distinguished from *welfare*. Welfare is the set of states of affairs, mostly external to the subject but not necessarily so, which contribute to his well-being. Thus to a person's welfare belongs his economic situation, his family relations, his professional situation, as well as his internal resources, in particular his health.

All these facts influence the person's well-being positively or negatively. In particular, the person's happiness with life is influenced, positively or negatively, by this collection of facts. It will be my concern to study the nature of this relation and in particular analyze the contents of the concept of happiness.

1. THE FRAMEWORK FOR QUALITY OF LIFE: HUMAN WELFARE

A human being always lives in some kind of environment. This environment has many parts. Firstly, there is a physical environment, a landscape with its natural resources and its climate. Secondly, there is a cultural environment, a society with its laws and regulations, with its political system, its customs and other cultural expressions. Thirdly, there is a close psychosocial environment consisting of relatives, friends and co-workers.

This environment influences our lives in many different ways, also when seen from a logical point of view. The environment is the *platform for our actions*, it gives us the opportunity to indulge in various activities. These opportunities vary enormously in different parts of the world. Greenland, for instance, provides the opportunity to go out hunting and fishing but it gives very little opportunity for doing agriculture. Southern Europe makes it

35

L. Nordenfelt (ed.), Concepts and Measurement of Quality of Life in Health Care, 35–57.
© 1994 Kluwer Academic Publishers. Printed in the Netherlands.

possible to cultivate grapes and citrus fruits, but that part of the world is not suitable for raising reindeer. British society gives the opportunity to express one's opinions and establish a variety of political parties. This is currently not the case in Cuba or North Korea. We can easily multiply such examples from many sectors of human life.

While being a platform for action the environment also sets the *limits for our goals in life*. We cannot seriously want to achieve anything in every environment. We cannot create a fleet of ships in Hungary, we cannot build a capital on the slopes of Mount Everest, etc. We cannot change our fellow human beings however much we wish. We can probably not make a Nobel laureate of our son or daughter.

The environment then is both a platform for action and a basis for our setting of goals. These are two principal ways in which a human being can be influenced. The person's platform can be transformed with direct consequences for his or her possibilities of action, and in realizing this the person may also have to change his or her goals in life, which has further implications for the scope of activity. But the environment can also influence us in a very *direct*, physiological way. This is very obvious with some physical aspects of the environment. We are directly influenced by the climate, for instance, in terms of wetness and cold, and by the chemicals of the water and the air. But there is also a direct influence from our cultural and psychosocial environment in terms of stimulation and stress.

This whole many-faceted environment influences us physically and mentally, both in the short and the long run. It affects in particular our well-being. The environment is an extremely important, although not the only, foundation for our well- or ill-being.

Thus we should distinguish between the external conditions of well-being and the well-being itself. (By a condition of well-being I mean a condition which in general contributes to a person's well-being. The condition need not be strictly necessary.) A state of well-being is created, affected and annihilated by different combinations of external conditions. To the effect that such a combination of external states contributes positively to a person's well-being we call it a state of *welfare*. To the effect that it contributes negatively to the well-being of a person, or even creates ill-being, I call it *illfare*. In what follows I shall mostly refer to the positive case.

This characterization must now be supplemented. There is an important domain between welfare, as I have so far described it, and well-being. There are not only external conditions of well-being but also a series of internal conditions. It is not only the external world that affects our well-being. We are also to a great degree influenced by our own physical and mental *constitution*: how we are constituted in physical and "mental" elements, our physical and mental strength, our health and our character, as well as our inclinations and interests. Our *inner properties* are certainly continuously affected by the external environment, but there is an important constitutional basis that the environment cannot affect in any other sense than that it can annihilate it.

Let us coin the expression *inner welfare* for that combination of inner properties which lead to or positively affect our well-being. The opposite could then be called *inner illfare*.

Still this is not sufficient. There is an area which in a sense is a product of the outer and inner conditions, and which to a great degree affects our well-being. This is *human activity* itself. If we reflect for a moment we can see that the most typical way in which the outer and inner conditions influence us is through our own activity. In order to be affected by some state of affairs we must very often act upon it or react to it.

In order for economic resources to lead to well-being they must be *utilized*. In order to get crops or plants, in general, we must go out and cultivate the earth. In order for a tool to be appreciated it must be used. We must use our weapons or fishing rods in order to get meat and fish. In short, we must work in order to get food, and in today's monetarian society we must work in order to get hold of almost all kinds of utilities. In all these cases the external world influences us through our taking advantage of it or reacting to it.

It is only in rather special cases that the external world affects our well-being directly, for instance the purely physical influence on our senses or on our body in general.

One can perhaps also say that the external world can affect us, simply by our *realizing* that it is there. One can simply be happy about the fact that one owns the sum of 100,000 pounds. One can feel grief at the famine in Africa. Neither of the these facts need directly or physically influence us. In spite of this we can feel both happiness and grief because of them. Perhaps we can also here say that this is a case of influence without the involvement of human activity. This, however, is a truth with modification. In order for influence to occur in this kind of case it is needed that the subject *reflects* upon the facts. It is as a result of this kind of reflection that he or she feels happiness or sorrow.

I can sum up in the following way. To human welfare belongs the external environment of the person, as well as his or her inner constitution and his or her own activities. All these things can affect the person's well- or ill-being. In addition, there is a series of both logical and empirical connections between these main categories. The environment can influence the constitution, the environment together with the person's constitution can influence the person's activities, and so on. I shall not, however, here further explore these connections.

Consider now the relation between the concepts of welfare and well-being. By the welfare of a particular person I mean the compound of things in his or her external and internal environment which together with his activities influence his well-being. A presupposition for an entity's belonging to the welfare of P is thus that it as a matter of fact contributes to the well-being of P. The set of states constituting P's welfare has thus not first been established in some independent way without any connection to P's well-being. The relation between the concepts of welfare and well-being is thereby *logical* and not empirical.

An immediate consequence of this is that the welfare of P need not be identical with the welfare of Q. The conditions for P's having a good time can be distinct from the conditions for Q's being well. Those things that make life good to a banker in New York are at least partly different from the things that make life reasonable to a housewife in Helsinki.

The welfare-classes of different people are thus partly different. But it is also clear that all of them have a common core. The conditions of pure survival are necessary elements among the conditions for well-being. Thus such conditions belong to the welfare of all people. Most probably some further conditions belong to the common kernel of welfare. It is, for instance, quite probable that a certain minimum of social relations have a place there.

2. TOWARDS AN ANALYSIS OF THE NOTION OF HAPPINESS

Welfare is the foundation for quality of life but is not the same as quality of life, as it is to be analyzed here. Quality of life resides in the well-being of a person and it can for most purposes — as I shall argue in the following — be identified with that subspecies of well-being which I call happiness with life.

My point of departure is that the concept of well-being covers the whole area of positive human experiences, from *sensations* to *emotions* and *moods*. It is customary in modern philosophical psychology to differentiate between these three categories. Sensations are those feelings which have a clear bodily location. Typical examples of sensations are pains and itches.

When one feels pain one feels it at a particular place on the body, for example one's knee. When there is an itch it occurs at a particular place, for instance behind one's ear. Emotions and moods do not have this bodily location. When one is in love (a case of emotion) one does not have a feeling which is situated at any particular part of the body. Nor does anguish (a case of mood) have a specific bodily site. (Both in love and anguish there may be certain sensations involved which are localizable. But none of these sensations can be identified with *the* love or *the* anguish.)

It is also important to differentiate between emotions and moods. The feature distinguishing between the two is that emotions are directed towards *objects*, normally outside the person him- or herself, whilst moods have no such direction and lack objects. The class of emotions is comprehensive, including love, hatred, joy, sorrow, hope, despair, etc. When one is in love one is in love with *somebody*. Similarly, one is sorry about *something* and one hopes for *something*. (For a more detailed analysis, see [2], pp 52-75 and [6], pp. 81-86.) As examples of moods, on the other hand, we find calmness, (certain instances of) depression and anguish. These lack objects and they are not located in any specific part of the body. They are feelings, which, if at all localizable to the body, are properties of the body as a whole. (In order to avoid misunderstandings I must here underline that these theses about the location of emotions and moods are not meant as theses concerning the neurophysiology or biology of feelings. They concern the phenomenology of feelings; they concern how we as ordinary human beings look upon our feelings and how as a result of this we use the words of feelings in our language. To say, for instance, that we do not locate a feeling of harmony in a particular part of the body is therefore completely compatible with a scientific hypothesis to the effect that there are certain localizable neurophysiological and endocrine processes which constitute preconditions of a feeling of harmony.)

Can the different emotions be directed to any kinds of objects? An analysis of the emotion-concepts shows that there are quite clear limitations. One can in fact give a salient demarcation of the kind of thing that can function as object of a specific emotion. For instance, one cannot hate a person unless one believes that this person has performed terrible deeds which are highly threatening to oneself or some of one's interests. One cannot despise a person unless one believes that he or she has behaved in a cowardly or otherwise morally reprehensible way. One cannot hope for something unless one believes that what one hopes for is something good and that there is a chance of getting it.

These statements are not conclusions from empirical observations. They are derived from the logical grammar of the concepts involved. A person who uses the terms 'hate', 'despise' and 'hope' in other ways cannot be understood unless he or she gives completely new definitions of these terms.

With these distinctions as a background we can now return to the different species of well-being. How should they be characterised according to this division? As a candidate for a sensation we can find at least one, viz. sensual pleasure. The receptors of the senses, such as smell, taste and the tactual sense, can give us sensations of pleasure. In our consciousness we would also locate these sensations in the relevant parts of the body, in the mouth, the nose or some part of the skin.

Among the moods we can find several kinds of well-being. To these belong for instance calmness, peace of mind and harmony. These feelings have no particular physical place. We feel neither peace nor harmony in the legs, the stomach, the heart or the brain.

What then about happiness? It is not a sensation. Happiness is clearly not located in a particular part of the body. One does not feel happiness in one's toes or in one's stomach. It is equally clear that happiness can take objects. One can be happy *about* something. One can be happy about one's progress, one can be happy about one's family life and even about life in general. Happiness in all these cases, therefore, is an emotion. (It could be argued that there is also a mood-sense of happiness. This would be the case when we say of a person that he or she is simply happy, without having anything to be happy about. I do not wish dogmatically to rule out this case. I only wish to stress that there may be an object of a person's happiness without him or her being aware of the object. Many cases of "just being happy" could therefore, be analysed in terms of unconscious objects.)

Happiness has a special position among the species of well-being. This is, I think, the reason why happiness should be the primary candidate for being identified with quality of life. The concept of happiness is in one sense the most general of the concepts of well-being. In a way it can be said to incorporate the other species. It can do so through its nature of being an emotion, i.e., by being a feeling directed towards an object. The object of the emotion of happiness can be of many kinds; and for this particular reasoning it is crucial to note that the object can be *another feeling*, for instance a feeling of the well-being kind. One can be happy about being at ease or one can be happy about an experience of pleasure.

Another way of expressing this is to say that happiness is a species of *well-being of the second order*. Happiness is a consequence of one's reflecting upon one's life. One observes and reflects upon some phenomenon in life and as a result one feels happy about it. Among the things that one reflects upon are the sensations and moods of one's well-being. And through this assessment there arises an emotion of happiness (or unhappiness).

A further related fact about happiness which puts it in a privileged position is that happiness can take a person's *whole life* as its object. One can be happy or unhappy about life as a whole. This then constitutes the result of a balanced reflection upon all parts of one's life that one considers relevant. Thus happiness has a further suitable property for becoming identified with general quality of life.

3. THE BASIC NOTION OF HAPPINESS

Is there then an abstract way of characterizing the object in the logical grammar of happiness? What can we be happy about? A classical answer to this runs in the following direction: Happiness is conceptually connected to the wants and goals of human beings. (For various formulations of this answer, see [3], [8] and [9].) One is happy about the fact that one's wishes and goals are realized or are becoming realized. One is happy about being able to do what one intends to do; one can be happy about an academic achievement or an achievement in sport; or one can be happy about some external event which contributes to the realisation of some of one's goals. One may inherit a fortune which makes a journey to Australia possible.

If one's life *as a whole* is characterized by the fact that one's most important goals are fulfilled or are in the process of becoming fulfilled, then this life is — with great probability — a life of great happiness. It is important to emphasize that the goals talked about here need neither be conscious to the individual nor be the result of a personal achievement or even constitute a change. One of our most important goals is to maintain the *status quo*, to keep our nearest and dearest with us, to keep our jobs and in general maintain our most fundamental conditions in life.

This observation about the relation between happiness and the realization of our conscious or unconscious wants gives us immediately a fruitful suggestion for judging the role of external conditions in the measurement of quality of life. Let us first look at the matter from an individual point of view. The external states of affairs which have directly to do with *my* happiness are those which contribute to — or prevent — the realization of *my* goals. If I have a professional role which entails great freedom of action and many resources, then this contributes to my happiness to a great degree.

On the other hand, a crisis in my family which threatens my fundamental security and binds all my energy prevents the realisation of my plans and reduces my happiness greatly.

Now, different people partly have different life-plans. Some of us have very ambitious and expansive goals; others, the more cautious people, set their goals on a lower level and require little from their lives. Some people wish to develop some aspects of their personality; others strive to develop other aspects. Thus the external conditions for happiness can vary enormously between individual persons. Consider, for instance, some extreme pairs: a monk and a disc-jockey, a mathematician and an athlete, a university president and a photographer's model, an alpinist and an archivist.

What does an observation like this tell us? It says among other things that we can never reliably characterize the happiness of a particular person just by describing certain parts of this person's life-situation unless we know about the relation between this situation and his or her wants and goals in life. There is therefore no valid and generally applicable instrument for the measurement of quality of life devoid of a systematic connection with the goal-setting of the individuals whose quality of life is to be measured.

With this reasoning as a background I shall now give a preliminary characterization of the concept of happiness:

> P is happy with his life as a whole, if and only if P wants his conditions in life to be just as he finds them to be.

A more formal and abstract way of expressing this thought is the following:

> P is completely happy with his life as a whole, if and only if,
> (i) P wants at t that $(x1...xn)$ shall be the case at t,
> (ii) $(x1...xn)$ constitutes the totality of P's wants at t,
> (iii) P finds at t that $(x1...xn)$ is the case.

Another more general way of expressing this intuition is to say that there is an equilibrium between P's wants and the reality as P finds it to be. The concept of happiness which I characterize here could therefore be called *happiness as equilibrium*.

It follows from this characterization that happiness must be a dimensional concept. P can be more or less happy with life according to the degree of agreement between the state of the world as P sees it, and P's wants. Moreover P can be completely happy with life only if his conditions in life are exactly as he wants them to be. Similarly, P is completely unhappy with life, only if nothing in P's life is at all as P wants it to be. There is then a continuum from complete happiness to complete unhappiness. This continuum of happiness must be distinguished from any particular state of happiness.

I shall say that the opposition between happiness and unhappiness is of a *contradictory* kind. This means that the continuum can be divided into two mutually exclusive parts, one part of happiness and another part of unhappiness. Later on I shall try to characterize the point at which happiness and unhappiness meet. I shall suggest a notion of *minimal happiness* based on the notion of a *high priority want*.

To the global notion of happiness with life corresponds a molecular notion of happiness with a particular fact. P can, for instance, be happy with the fact that he or she has passed an exam. P is happy because he or she had wanted to pass this exam. In general, P is happy with every fact which constitutes the satisfaction of a want of P's. In a way global happiness with life constitutes the sum of "molecular happinesses" with particular facts.

This sum, however, cannot be derived in a simple arithmetic way. Our general happiness or unhappiness concerning life is dependent, not so much on the quantity of matters that we are happy about, but on what kinds of matters we are happy about, in particular on what we consider to be important in life. To most of us it is more important to become a parent than have a nice day out in the forest. The father's happiness about his newborn baby influences his general happiness much more than his happiness about the beautiful weather.

3.1. The reference of happiness to different points in time

I said that P is happy now if the state of the world is as P wishes it to be. This reference to the present time is important and deserves further comment. It is plausible to think in the following way: if happiness is connected with the fact that wants have been satisfied, then happiness ought to be connected with the past. That person is happy, one should have thought, whose wants in the past, including the most recent past, have been realised. I can easily show that this idea is not sufficient to explain the nature of happiness. Consider the following case:

A small boy has for a long time wanted to have an electric train. He has wanted this intensively and told his parents about his desire. He is given this kind of toy as a Christmas present. He is certainly very happy. However, after a short while he becomes terribly bored. He finds that there is in fact so little that he can do with the toy. The happiness has very quickly been transformed into boredom.

What has happened if we wish to describe this situation more abstractly? It is true that the boy has had a want from the past satisfied. However, this want no longer exists. At the present moment there is no want for the train; therefore the presence of the train cannot satisfy a want on the part of the boy; therefore it cannot contribute to happiness.

The important and indeed well-known fact to be illustrated by this example is that the satisfaction of wants can be followed by such emotions as disappointment, boredom and regret. Therefore the wants whose satisfaction should constitute happiness must refer to the present time.

The reference to the present time solves a further problem which is often mentioned in dissertations about happiness. A person can say the following: I am happy about a present that I have received, but I had never expected to receive it and I had never wanted it. No, it can very well happen that something new occurs in one's life; one may not even have known of its existence. Therefore one cannot have wanted it in the past. But when it occurs, one may quite strongly want to hold on to it; one simply likes it, as we say. Thus its existence at the present time contributes to one's happiness.

But what then about wants which are directed at the distant future? Do they have anything to do with one's happiness? Consider a young man who is planning his life. He plans to marry in ten years; he plans to complete an education which takes at least five years; after that he wishes to enter a long career as, say, a lawyer. In short he has many wants referring to the very distant future.

Per definition, these wants cannot be satisfied now. If they were they would not be wants directed to the future. But the fact that these wants are not satisfied now does not entail that the boy is unhappy now. This would of course be an absurd conclusion.

However, there is an important indirect way in which future-related wants also affect present happiness. If the present conditions are such that they do not favor the realization in the future of P's future-related wants, or if they even prevent this, then P's happiness will be affected negatively. P is happy now, only if he believes that the *present prospects* of the realization of his future-related wants are good. These good prospects are in a way conditions existing *now* which P wants to be the case.

It can be argued that P may also want his *past* to have been different than it was. P may, for instance, be unhappy *now* because he never established a close relationship with his father. Since his father is dead, this is a state of affairs that cannot be improved. Unhappiness related to an unalterable past can therefore only be removed by a change in one's *perception* of the relevant past and the *importance* which one attaches to it. Both these notions will be discussed in subsequent sections.

3.2. The dependence of happiness on belief and knowledge

In order to be able to want something one must be a minimally intellectual creature. One cannot want to have a car unless one can imagine a car. One cannot want to take an exam if one does not know anything about exams and their relevance for certain professional careers.

This truth has immediate consequences for the concept of happiness that we are trying to reconstrue. I cannot be happy about x unless, at least, I believe that x is the case. The sources of this belief, however, can be of various kinds and various validity. Most importantly, they can be both true and false.

If P is happy about x, then P believes (or knows) that x exists or occurs and that this constitutes the satisfaction of a want of his. But as we said, this belief need not agree with reality; what it agrees with is P's *perception and awareness* of reality.

The concept of happiness to be construed thus is a *cognitive* concept. To be happy with life one must have a certain set of beliefs (or some stronger cognitive states, such as conviction or knowledge). This does not imply that the beliefs involved must be sophisticated. They can be very simple and be related only to immediately present persons and things. Thus, the theory clearly allows that small children and profoundly mentally retarded people can be happy. The theory would indeed be defective if this were not true.

3.3. On happiness and pleasure

What is the relation between happiness and other species of well-being? Let us face the classical question concerning the relation between pleasure and happiness. (For a thorough discussion of this problem area, see [1].)

Let me try to answer this question by considering what can be a *reason* for wanting something. Why do we want to have something. Why, in general, do we want something to be the case? We can give many answers to such why-questions. They can be different for different people. But there is a typical answer to such a why-question and this is the following: I want x because x gives me immediate pleasure. This is a way of terminating the series of why-questions. There is no point in asking further questions.

Psychological hedonism is a theory which maintains that all our wants refer ultimately to a state of pleasure of some kind. The pleasure need not be the immediate reason. The chain often has more than one link. It can have the following structure. I want to have a car at hand to be able to transport myself to the theatre. I want to get to the theatre to see an interesting drama. I want to see this drama for the sake of intellectual pleasure. This hierarchy of wants which terminates with the want for pleasure is indeed typical. (I do not, however, as the psychological hedonists do, consider it to be the only kind of hierarchy that there is.)

Pleasures are states of mind which are typically appreciated for their own sake. This does not, however, mean that pleasure is identical with happiness, nor that a person who experiences strong pleasure is automatically happy. A person is of course normally happy about his pleasure, or about his absence of pain, but there are cases where this need not be so.

Consider the case where the pleasure is a sign that something dangerous is going on. The pleasure involved in the taking of a drug, for instance. The addict may be conscious of the fact that after a while the pleasure will be gone and that the future suffering will be great. Hence, although at a particular moment he may experience intense pleasure, he may at the same time be deeply unhappy.

Conversely, pain and suffering are states of mind which are typically unwanted. Normally, a person in great pain is unhappy about his or her state of mind. But again this need not be so. The pain or suffering may be a sign that something positive is to be expected. A surgical operation may be very painful.

However, if the patient believes that the operation is an effective measure, and that he or she will soon be healthy, then the pain is highly endurable and can be coexistent with great happiness.

3.4. On the minimal quality and quantity of wants

Can a person not have too low aspirations, or too small a set of wants? Can the want-equilibrium criterion work in the minimal case? Is it reasonable to say of a man who does not want to do anything else than to lie on a sofa, licking a lollipop, that he is happy, even completely happy, when this want of his is satisfied?

Given some necessary specifications of the example I shall say that this man can be happy in the want-equilibrium sense and that it is also reasonable to say so. What we must presuppose in this example is, of course, that this man now and again wants to eat, drink and sleep and that these wants are satisfied. If this were not so, the man would soon become hungry, thirsty and tired, which eventually would create great suffering on his part. We must also assume that the person does not frequently become bored, that he does not in fact quite often wish he were somewhere else doing something useful. In short, we must check the truthfulness of the quite extreme statement that the man really does not want to do anything more than he claims.

In most cases, when one reflects upon it, it is simply not true to say of a person that he or she only has a limited set of very trivial wants. The psychologically more plausible picture of the man on the sofa is that he is terribly bored with life in general and desperately wants to do something else, but he may be unable to change his situation for various psychological or circumstantial reasons. In such a case the man is, of course, profoundly an unhappy person.

Having said this we must allow for the extreme case of a person who is not bored and who genuinely does not want to do anything else. This would presumably be a person with a very low degree of intelligence and emotional sensitivity. But it would be wrong to deny that he has happiness in the want-equilibrium sense of the word. His happiness in the sense of *richness* (to be introduced below) would, however, be low.

3.5. Being happy and feeling happy

In introducing the notion of happiness I described it as a *feeling*, belonging to the subspecies of emotions. Let me now qualify this, and distinguish between happiness as a *disposition for feelings* and happiness as an *occurrent feeling*.

It is obvious that we often talk about "feeling happy". One can be said to feel happy with life as well as feel happy with more particular conditions in life. Happiness as a feeling can be more or less intense, more or less frequent and more or less enduring.

There is, however, a point in distinguishing between *feeling* happy and *being* happy, where the latter refers to happiness as a disposition not necessarily entailing any particular experience on the part of the subject. The two kinds of states are certainly analytically connected. From the fact that P feels happy follows that P is happy. And from the fact that P is happy follows that P is disposed to feel happy.

It is clear that the dispositional mode of speech is also frequent. We often talk about a person being happy with many different things in life, without thereby entailing that this person has distinct feelings corresponding to all the facts he or she is happy with. Moreover, the person need not at a particular moment have any feeling whatsoever. He or she may be asleep or may be quite involved in sorting out some trivial matters in life.

There is, however, always a connection to feeelings. If a person is happy in the dispositional sense, then he or she will feel happy under certain conditions. A particularly favorable condition for that kind of feeling to occur is that the subject *pays attention* to the state of affairs that he or she is happy about.

3.6. On different degrees of happiness

We have said that happiness can be viewed as a dimension. A person can be more or less happy. But how should we understand this dimension? And what determines a person's degree of happiness?

Since happiness is conceptually connected to the agreement or disagreement between a person's wants and reality as he or she finds it, it is tempting to relate happiness to the number of wants that have been realized. Suppose that P has 100 wants and that 90 of them are fulfilled. Suppose, on the other hand, that Q has equally many wants but that only 10 of them are fulfilled. According to a simple arithmetical calculus, P's happiness ought to be 9/10 of the possible total happiness, whereas Q only reaches 1/10 of this degree of happiness. Hence, P must be much happier than Q.

A moment's reflection shows that this must be a caricature of reasoning. It is a caricature for a number of reasons. One important such reason has to do with the idea of a want-unit. What is *one* want as opposed to a number of wants?

Many of our wants are logically related in hierarchies. One may want to have F in order to get G. G in turn may be wanted for a further end, and so on until we reach an ultimate end wanted for its own sake. How are the wants to be counted here? Shall we say that we have just one basic want? Or is every member of the hierarchy a separate want? Both alternatives involve difficulties. If we were to count only the basic abstract wants, we might end up with quite few entities (for instance, a want to live a long and successful professional life, a want to live in one's own country with one's family) which may not differ so much between individuals. The other alternative of counting all elements in every chain leading up to the fulfilment of the basic wants is, at least in practice, an impossible alternative. For one thing, such a chain is infinitely divisible. If one wants to walk to the grocery store, then — it may be claimed — one wants also to walk to each point on the way to the grocery store.

Consider a further difficulty. The two wants, to want F and to want G, can always be construed as the single complex want, to want F and G. Thus any number of contemporaneous wants can always be reduced to one. (Observe that the wants in order to be collapsed to one need not *refer* to the same time. One can have the single complex want of fulfilling F at t1 and G at t2.)

I have said enough to introduce some great problems into the project of using arithmetics in the characterization of a person's want-structure. I shall not pursue this further. It is not necessary for my main reasoning, which will completely avoid the counting of wants.

I shall instead introduce the idea of *priority*. I shall say that there are wants of higher and lower priority. And it is the degree of priority which determines whether great happiness will result from satisfaction of a want or not.

Some of our wants have a vital importance to us. To most of us it is very important that our family is well and is successful. Our own health is also of great importance. So is the fact that our professional situation is all right and perhaps also that the political situation is tolerable. That these facts are the case has considerably higher priority than most other things we wish to do or have at a particular moment.

Thus there must be a scale of priority or importance, along which we can rank our wants. This ranking is practically never explicit, nor is it particularly clear when we try to visualize it. And it is absolutely certain that there is no "naturally" given cardinal order for these wants. We cannot say that it is five times more important that our children are alive than that our own health is in order.

But what then can we say about this scale of priority that certainly exists with every human being? First, how can we know, or what is the criterion of the fact, that a certain want of P's has a higher priority than another want? I shall suggest the following characterization:

> Want x has to P a higher priority than want y, if and only if P in a choice between x and y, where both cannot be realized, would prefer x.

With this formulation I can keep the connection to my basic analysis of the concept of happiness. I said that P is happy about life, if and only if P wants the conditions of life to be exactly as he finds them to be. I can now say that P is more happy about a situation which contains x than about one which contains y and explain it by simply saying that P *prefers* the situation with x to the situation with y.

Given this explication we seem to have an intrapersonal instrument for comparison. We can understand how to analyse and also in principle how to get to know that a person is happier now than he was before. But do we now have an instrument for a comparison between different people? How shall we explicate the idea that Brown is happier than Smith, or for that matter that Brown is equally as happy as Smith?

The analysis that we have given will allow us to do this only under very specific circumstances. Assume that Brown and Smith have exactly the same profile of wants. That is, they have exactly the same wants and their priorities among the wants are identical. Thus we know that, if Brown prefers x, Smith must also do so. (We must here presuppose that x and y denote total situations.) Assume now that Brown is in x and Smith is in y. Then it follows that Brown is happier than Smith.

A pragmatic method for interpersonal comparison concerning happiness would then be to describe to the people involved their respective life-situations. To make it as simple as possible: We compare the happiness of two persons A and B. We describe A's situation meticulously to B and the other way round. It appears that A prefers his own situation to B's and that B prefers A's situation to his own. Then we have reason to say that A is happier than B.

Having indicated this procedure for interpersonal comparison, we must notice two great difficulties in practice. "Preference" in this context is an ideal concept. We must be talking about an ideal situation of choice, where the individual has complete self-knowledge and can foresee such things as risk of disappointment, boredom etc. It is certainly true that people normally lack this self-knowledge in actual situations of choice.

It is also important to stress that one has to compare people's total situations in order to be sure that the result mirrors their state of happiness. P who is in x need not be happier than Q who is in y (even if both of them prefer x to y) if x and y do not cover total life-situations. If x and y only affect some part of their lives, for instance professional life and state of health, then it is always possible that something unhappy has happened to P within some other sphere of his life. P can have lost some close relative, a fact which can have created deep sorrow in him.

In general it is easy to go wrong when one dreams about another person's life situation and prefers it to one's own. It is easy to prefer the life of a shipping magnate to one's own. But one has then to remember that to his life situation belongs not only his money and his yachts but also his diseases and his "love problems".

But how should we then treat all those cases where the goal profiles are different between people and where a comparison between two persons' life situations does not give the result that both agree on what situation to prefer? Can we then ever say that a person is happier than someone else?

My general conclusion is that there are many cases where the happiness of people are *incommensurable* magnitudes. We can in these cases simply not — not only for practical but also for theoretical reasons — say that one is happier than the other one.

There is one important exception to this, though. This is the case where P finds his situation *unacceptable*, while Q finds his acceptable. In this case Q is clearly happier than P. And we can say this even if Q were to prefer P's situation to his own, and P were to prefer his to Q's situation in life.

By introducing the notion of acceptability I wish to indicate where the line is between happiness and unhappiness on the happiness scale. I suggest here that with every human being there is a level which marks the transition from happiness to unhappiness. Below this level the situation is so far from satisfactory that it is not acceptable to the subject. The subject is unhappy. Just above this level the situation is acceptable; he or she is *minimally* happy.

Preliminarily we can characterize this line in the following way: In order for P to be at least minimally happy, then all those conditions which have a high priority for P, in an absolute sense of the word, must be materialized. Where this line goes in any concrete sense must vary much between different people. People have different temperaments and traits of character. The impatient and spoilt person is such that he or she becomes unhappy for the most trivial reasons. To such a person then almost every want has a high priority. The patient or stoic person, on the other hand, is such that he or she can meet most adversities without falling below the level. To this person very few matters in life have a high priority.

This observation about the dependence of happiness on how we make our priorities, contains a key to happiness which we have so far not recognized. To influence the happiness of people means not only to try to realize states of affairs in their external or internal situation. It can equally much entail influencing their profiles of wants. That person who has a low profile, the person with the smallest number of wants with a high priority, has, in one sense, the greatest chance to become happy.

3.7. A further dimension of happiness: the dimension of richness

Consider a boy who has lived all his life in simple and unpretentious circumstances in the Highlands of Scotland. He has been entirely content with his lot; he gets along well with his family, he appreciates the wild landscape and he enjoys the sometimes hard struggle for life up there. Thus, for a long time he has been completely happy in our want-equilibrium sense.

One day he and his family are visited by a tourist who happens to be a famous musician. The tourist is attracted by the place and settles there permanently. Partly to earn his living he starts teaching the young boy to play the violin. He then discovers that the boy has a remarkable talent for music and that he very soon develops a proficiency in playing. This completely changes the boy's life. A whole new world has been opened to him and he acquires a set of completely new goals in life. To put things in my technical terms, he has acquired a multitude of new wants, which he didn't have before, and he is in the process of satisfying them.

The boy was, as we said, completely happy with life before the musician arrived. But how should we then express the positive change that has now happened to him? Let us here presuppose that it is a positive change. One can, of course, imagine the case where the appearance of the musician disturbs the idyll in the family. The boy's parents may become envious or even jealous; they may want to get rid of the musician and thereby ruin the boy's life.

But the case that we shall consider is instead the following: the boy has all his previous wants satisfied; in addition he now has a number of further wants; some of these, or even all of them, are being satisfied. In quasi-numerical terms, the boy may now have 20 out of 20 wants satisfied, instead of 10 out of 10 wants satisfied.

In order to characterize this case I shall now introduce a further dimension of happiness, which I shall call the *dimension of richness*. The Scottish boy is now happier than before in the dimension of richness, but strictly speaking equally as happy as before along the equilibrium dimension.

Richness does not solely concern an increase in the number of wants. It can also entail the case where some modest wants are replaced by a set of more *ambitious* wants. Let me exemplify such a case with the young man P who starts his professional life with very low expectations and ambitions. He wishes to become a clerk in a bank like his father and starts working in the local branch of the national bank. His life fulfils all his modest expectations and ambitions. Thus P is completely happy in the equilibrium sense. Later, however, he is persuaded to set his goals higher. He starts an advanced course in banking and soon reaches a top position in his bank. As a result of this change he has become, as he also himself claims, much happier. Still, in the want-equilibrium sense, he was completely happy already from the beginning.

This change of a person's state of happiness cannot be explained in terms of an increase of the absolute number of satisfied wants. Still, it can be characterized within our general conception. When the Scottish boy has become a musician we let him compare his present life as a whole with his previous life. He then finds that he prefers the new life to the old life. Similarly, P prefers his life as a bank manager to his life as a clerk. Ex post facto he can then say that the new life is in greater agreement with how he wants his life to be.

(This reasoning presupposes, however, that the order of preference does not change over time. P can at t1 prefer the life P now has to the one that he had at t0. This does not necessarily entail that P is happier at t1 than at t0. P's order of preference at t0 can have been different than at t1, so that P at t0 in fact would have preferred the life at t0 to the one that he was going to have at t1.)

A way of concluding this section is to say that people have a set of potential wants which can become actualized depending on the way the people are informed. As long as a person still has an unactualized set of potential wants he or she can become happier in the richness sense of the word.

4. SOME PROPERTIES OF THE RICHNESS DIMENSION

4.1. The upper limits of richness

Can there be anything such as complete happiness in the dimension of richness? As we can see, there is no simple analogue to the mathematical algorithm that can be used concerning happiness in the want-equilibrium sense. In the latter case there is something that could be called complete happiness. Such happiness is obtained when all the person's wants have been realized.

But in the richness case we can see no such a priori limit. Happiness can therefore in principle be extended indefinitely along this dimension.

The limit that there must be is of an empirical kind. There is a physical and psychological limit to what people can grasp intellectually and emotionally; as a consequence there is a limit to what they can take an interest in and form wants about. Moreover, and what is more confining, there is a limit to what they can do and achieve.

Furthermore the limit varies with individuals. Different human beings have very different abilities to grasp reality, to take an interest in reality and to realize advanced plans in life. There is a personal upper limit for how happy P can become in the richness sense, which is different from what Q can achieve. In contrast to the want-equilibrium sense, according to which everybody can become equally happy, there is a constitutional difference in people's ability to become happy in the richness sense.

4.2. Does a high degree of happiness in the richness sense require complete happiness in the want-equilibrium sense?

Siri Naess ([4], p.20) says in a criticism of the simple want-equilibrium theory:

> A person who has 90 out of 100 wants satisfied can have a much better time than the person who has all his 10 wants satisfied.

We must first observe that the person who has 90 out of 100 wants satisfied can be and perhaps typically is *un*happier than the one with all his or her 10 wants satisfied. There is a great risk that the former has, among the ten which are not satisfied, some high-priority want. Thus he or she is, per definition, unhappy with life.

Assume, however, that all the ten of the 100 unsatisfied wants are minor. The person is, then, also according to the want-equilibrium criterion, a happy person. We can imagine him or her to be an extremely vital and active person who has a great number of projects and aspirations, both professionally and for his or her leisure time. In fact the person wants to do so many things that he or she has no time to realise everything. This person lives a rich life, and can be very happy in the richness sense, without necessarily being completely happy in the want-equilibrium sense.

In connection with this observation we may notice that, given a psychologically plausible assumption, the non-satisfaction of a certain number of wants is compatible with complete happiness in the want-equilibrium sense. Consider the person who wants his or her life to be an adventure or a struggle, in general to be somewhat unpredictable. In more theoretical terms this may mean that this person has a higher-order want which is such that he or she does not want all first-order wants to be continuously satisfied. The person may indeed want certain preoblems to arise in order to remain alert and highly aware. This second order may indeed have a high priority. Thus, if the non-satisfied ten wants, mentioned in the quotation from Naess, are within the scope of this second-order want, then, slightly paradoxically, our vital person can achieve complete happiness in the want-equilibrium sense.

5. THE THEORY OF HAPPINESS SUMMARIZED

Let me here summarize the conceptual framework which I have introduced here.

The main components in this system of concepts are *welfare*, *well-being* and *happiness*. *Welfare* is distinct from *well-being*. The two phenomena are, however, seen as analytically related. A condition C belongs to the welfare of a person P, if and only if C contributes to the well-being of P to some degree. Thus an individual's welfare consists of a great variety of conditions, including conditions which are both external and internal to the individual. For instance, the person's health belongs to his welfare. So do his own activities.

Happiness is viewed as a species of *well-being*. Happiness is either an emotion or a disposition for an emotion. Happiness with life is characterized in the following way: P is happy with life as a whole, if and only if P wants his conditions in life to be just as P finds them to be.

It is argued that happiness has a privileged position among the species of well-being. Happiness can in a sense encompass the other elements of well-being. According to the suggested characterization happiness is a consequence of one's reflecting upon matters in life. Among the things that one reflects upon are the sensations and moods of one's well-being.A further important fact about happiness is that it can take a person's whole life as its object. One can be happy or unhappy about life as a whole.This then constitutes the result of a balanced reflection upon all parts of one's life that one considers relevant.

For most purposes, I argued here, a person's *quality of life* should be identified with his or her *happiness with life*. Quality of life is under this interpretation a subjectivist notion.

From the foregoing analysis of happiness it follows that happiness is a cognitive concept. Happiness presupposes knowledge and belief of a particular kind. It also follows that happiness is distinct from pleasure as well as from other sensations and moods which are positively evaluated by the subject.

Furthermore, happiness is a dimensional concept. One can be happy in different degrees. There is in fact a scale from maximal happiness to maximal unhappiness. I have discussed different ways of characterizing this scale, one quantitative and one qualitative. I have dismissed the quantitative idea, viz. the one that entails that a person's degree of happiness is dependent on the number of his or her satisfied wants. Instead, I proposed the idea that a person's degree of happiness is dependent on whether the person's satisfied wants have a high ranking in his or her hierarchy of priorities.

In my final section I noted that the initial definition of happiness remains problematic because a person's set of wants is dynamic. New wants can be introduced and a person's happiness can rise through the expansion (and the subsequent satisfaction) of his or her set of wants. The concept of *happiness-as-equilibrium* (as given in the initial definition) is thereby supplemented by *happiness-as-richness*.

NOTE

The theory proposed here is more extensively discussed and defended in [7].

BIBLIOGRAPHY

[1]   Gosling, J.C.B.: 1969, *Pleasure and Desire*, Clarendon Press, Oxford.

[2]   Kenny, A.: 1963, *Action, Emotion and Will*, Oxford University Press, Oxford.

[3]   McGill, V.J.: 1967, *The Idea of Happiness*, Fredrick A. Praeger Publishers, New York.

[4]   Naess, S.: 1979, *Livskvalitet: Om å ha det godt i byen og på landet*, Institute of Applied Social Research, Oslo.

[5]   Naess, S.: 1987, *Quality of Life Research: Concepts, Methods and Applications*, Institute of Applied Social Research, Oslo.

[6]   Nordenfelt, L.: 1987, *On the Nature of Health*, D. Reidel Publishing Company, Dordrecht.

[7]   Nordenfelt, L.: 1993, *Quality of Life, Health and Happiness*, Avebury, Aldershot, U.K.

[8]   Tatarkiewicz, W.: 1986, *Analysis of Happiness*, Martijnus Nijhoff, The Hague.

[9]   Telfer, E.: 1980, *Happiness*, MacMillan Press, London.

PAOLO CATTORINI AND ROBERTO MORDACCI

HAPPINESS, LIFE AND QUALITY OF LIFE: A COMMENTARY ON NORDENFELT'S 'TOWARDS A THEORY OF HAPPINESS'

Nordenfelt's thesis is very clear. His conceptual framework, in which *quality of life* and *happiness-with-life* are identified, is rich and well articulated. It can be summarized as follow:

Happiness-with-life is a positive human experience (it is, in fact, "a species of well-being of the second order" p. 40) with the *whole life as an object*. It is important to note that, in order for an experience to be positive, the object must be thought to be *something good* (p. 40). Happiness is clearly dependent on the goal-setting of the individuals, so that it is impossible to give a description of the emotional condition of a person simply by describing his external states of affair: in fact, this states, according to Nordenfelt, can be defined as *welfare* if, and only if, they contribute to that person's *well-being* (the relation is logical, not empirical). Happiness is then defined as an *emotion* deriving from the equilibrium between a person's wants and his conditions in life as he perceives them (*happiness as equilibrium*). Happiness has also a *richness* dimension, depending on the expansion of the set of wants that are satisfied. Nordenfelt concludes that happiness, and therefore quality of life, are subjectivist notions, whose measurement is very difficult, if not impossible.

The main line of the argument leads to the conclusion that "there are many cases where the happiness of people are *incommensurable* magnitudes" (p. 51), except when a subject finds his situation in life so poor that it is *unacceptable* for him: then he is clearly *unhappy* and therefore he has a *bad* quality of life. Anyway, the only way to know anything about a person's quality of life is to ask him (maybe through a complex set of questions) whether he is (or feels) happy with his life.

We think that his analyses raise some difficulties beyond the subjectivist theory of happiness, i.e., their direct application to the issue of the Quality of Life (QL). These are, to our opinion, the most problematic points:

1. First, we think that a better argument is needed for the identification of QL with a subjectivist notion of happiness-with-life. Nordenfelt does not offer any clear definition of what he means by *quality of life*, so that happiness and QL seem to be simply synonymous: this should be developed further. As will become clear in the following, we think that what is at stake in a QL judgment is not the same thing as in a judgment on a person's hap-

59

L. Nordenfelt (ed.), Concepts and Measurement of Quality of Life in Health Care, 59–62.
© 1994 Kluwer Academic Publishers. Printed in the Netherlands.

piness. Our first observation is simply that the identification Nordenfelt proposes is left unexplained.

2. In Nordenfelt's framework, as he clearly notices (p. 50), there are some difficulties in practice which make interpersonal comparison difficult: a person's "preference" about his own life is not the kind of judgment that can be assessed objectively. In fact, even from a subjectivist point of view, my judgment about the quality of *my life* is not simply my feeling more or less happy with it, but also with the perceived *worthiness* of my conditions of living. We think that this kind of judgment always claims, in the first instance, a certain universality, in the sense that it claims to be recognized by others on the basis of the common *experience of life as a possibility of good to be realized* (and therefore valued) [3]. It is for this reason that the enterprise of QL research could be resuscitated (if ever it is needed at all) only by a reference to "common sense", i.e., to a similarly shared experience of the world, as it has been proposed, for example, by Morreim ([4], pp. 226-227). Nordenfelt's criterion of *acceptability* is an attempt to express this claim to universality, but, as it remains completely subjective, it seems to us to be insufficient, particularly in relation to QL judgments.

Furthermore, QL judgments are important in situations where patients cannot express their preferences about their situation: it is very difficult to ask a terminally ill patient how he feels with his whole life (not to mention comatose patients). What we need in such circumstances is to know what kind of action on our side (medical intervention or simple assisting care, for example) best translates our concern for the patient, our effort, maybe, to *relieve* his unhappiness.

3. Characterizing happiness through the dimensions of *equilibrium* and *richness*, and defining it as an *emotion*, Nordenfelt almost completely dismisses the Aristotelian definition of happiness as *an end in itself*. Happiness itself is a want, or better it is *the* want. In the words of Aristotle, "we call final without qualification that which is always desirable in itself and never for the sake of something else. Now such a thing happiness, above all else, is held to be; for this we choose always for itself and never for the sake of something else" ([2], I, 7, 1097a-1097b). But can we say that the desire for happiness can ever be satisfied by an emotive equilibrium of our wants along with our perception of reality? Do we simply seek an *emotion* in searching for happiness? We think that there is a kind of radicality in the desire for happiness, that makes it something other than the search for an emotion (even if its object is one's whole life): happiness appears to be the emotional dimension of the full good of the person (what the classical tradition called

beatitudo or plenitudo essendi, or *eudaimonia*), which is the object of de-sire, assumed and translated by the will. In a QL judgment it is the *possi-bility of achieving the good of the person* that is investigated.

4. My feeling happy or unhappy with my life is indeed important in as-sessing my QL: the positive or negative appreciation of life in the emotions of the subject is a sort of pre-rational judgment on the QL that cannot be dis-regarded. But we think that in a QL judgment there is something more: a claim to assess one's concrete life situation in terms of the possibility of achieving the good. It is because life holds out a promise of good that it is *worthy* and can be said to have a *dignity* [1].

A subject who even defines his life as *unacceptable*, and therefore him-self as *unhappy*, may still (rationally or not) recognize the possibility, at least for his freedom, to have his unhappiness redeemed. And when the *claim* is exactly that this possibility is definitely and completely denied (a claim that we think is untenable), this is not simply an emotion, but pre-cisely a claim to objectively recognize one's situation. This is the reason why QL judgments are mostly made by others than the subject in question: a QL judgment is needed when we seek a reasonable way to decide if, and how much, the concrete conditions of a single life, at a certain moment, *contradict* the promising nature of life.

In conclusion, while recognizing the logical rigour of Nordenfelt's analy-sis of happiness, we think that it fails to address some important points in clarifying the nature of QL judgments. But we would like to advance a more radical question: Can QL ever be measured? Can the grade, the quality, and the quantity of good a life promises be measured? If we could do so, do we need it?

Life appears not as a simply biological fact but as an event in which the possibility for new meaning is always open, and in which the freedom of the individual is needed to search for that meaning [1]. This call cannot be ig-nored for the search is inevitable, unless one abandons oneself to an in-authentic life. To this extent, neither life nor its quality seems to be some-thing measurable, because the meaning calls for interpretation rather than for measurement.

BIBLIOGRAPHY

[1]   Angelini, G.: 1986, 'La vita: fatto o promessa?', *Rivista di Teologia Morale, 71*, 55-69.

[2]   Aristotle: 1954, *The Nicomachean Ethics*, trans. D. Ross, Oxford University Press, London.

[3]   Cattorini, P.: 1989, 'Qualità della vita negli ultimi istanti', *Medicina e Morale, 1989*, 2, 273-294.

[4]   Morreim, H.E.: 1992, 'The Impossibility and the Necessity of Quality of Life Research', *Bioethics, 6*, 218-232.

PER-ERIK LISS

ON NEED AND QUALITY OF LIFE[1]

The study of human beings and their environment normally involves models and theories of different kinds. In the heart of these models and theories we find concepts, the meaning of which also influences the practice these models and theories are means for studying. The central place of the concepts in the scientific tools, together with their significance for the organization and understanding of life, constitutes a strong reason for analysis. It is then important to consider not only the meaning of individual concepts, but also the logical and empirical relations between the concepts. An analysis of the logical relations between different concepts might, for instance, render empirical investigations of these relations superfluous.

Need, health, and quality of life are central concepts within the social and health services. In this essay I shall focus on the relation between the concepts of need and quality of life. The purpose is to describe and analyze the relation. Both the relation between the concepts of need and health, and that between the concepts of health and quality of life, are analyzed elsewhere.[2]

Consider the following statements on need and quality of life:

(1)　The satisfying of fundamental needs leads to, or results in, a high degree of quality of life;

(2)　the fundamental needs are satisfied if, and only if, a certain degree of quality of life is achieved;

(3)　a high degree of quality of life of P leads to, or results in, satisfied needs of P; and

(4)　P has a high degree of quality of life if, and only if, P's fundamental needs are satisfied.

The four statements are all reasonable, although perhaps to different degrees. They differ, however, in two important respects. First, (1) and (3) are statements about the causal relation between the phenomena need and quality of life. Assessing the correctness of these statements requires empirical investigation; (2) and (4) on the other hand are statements about the logical relation between the concepts of need and quality of life.

63

L. Nordenfelt (ed.), Concepts and Measurement of Quality of Life in Health Care, 63–78.
© 1994 *Kluwer Academic Publishers. Printed in the Netherlands.*

Assessing the correctness of these statements requires philosopical analysis. Second, (1) and (2) are primarily statements about need or the concept of need, while (3) and (4) state something primarily about quality of life or the concept of quality of life.

I shall only briefly consider the causal relation between need and quality of life. It is the conceptual relation, exemplified by the statements (2) and (4), on which I shall focus.

A common way of defining a concept is to specify necessary or sufficient conditions for the application of the term. Two concepts can be related mainly by the one being included, directly or indirectly, in the conditions of the other: (2) and (4) are examples of relations where one concept is included directly in the conditions of the other. In (2) the concept of need is defined in terms of quality of life, and in (4) the concept of quality of life is defined in terms of need.

NEED DEFINED IN TERMS OF QUALITY OF LIFE

In the extent literature on need or the concept of need we can discern two different traditions. In one, need is considered as something instrumental or goal-related. The statement "P needs X if X is necessary for realizing a certain goal" illustrates this view. In the second tradition need is considered as a tension or a disequilibrium in the organism. The term need here refers to drives or motivating forces. In this section I will restrict the discussion to the goal-related view of need. The view that need is a tension will be dealt with in the next section, where quality of life is defined in terms of need.

In a study on the concept of health care need [8] I sided with the goal-related tradition which relies on the following definition of the concept of need:

P has a need for X in situation S if, and only if, (i) there is a difference in S between the actual state of P and a goal G, and (ii) X is in S a necessary condition (means) for G.

According to this definition there is a need when condition (i) is fulfilled, and a need for X when both (i) and (ii) are fulfilled. The term need here refers to the difference — often a state of deficiency. But it is not unusual for the term to be used to refer to the *object* needed (X in the definition) — for instance, "love is a human need". Both these types of reference are used in this essay, although the focus is on the reference to the *difference*. It is important, however, that the distinction between the need (a difference or deficiency) and the object of need (a necessity for eliminating the difference or deficiency) is not neglected when analyzing the concept of need.[3]

The adherents of the goal-related tradition specify the concept somewhat differently, and the definition noted above is just one example. But they all take the view that need is something goal-related — x is needed when X is necessary in order to realize a *goal*. The goal component plays an important role in the concept. First, the goal *determines* the object of need (that is, the thing needed). The goal "to live as a farmer" generates a set of needs which partly differs from the set generated by the goal "to live as a lawyer in the city". Different goals generate different sets of needs. A clearly defined goal is therefore a prerequisite for a reasonable assessment of needs.

Second, the goal is the *justifying* component in the need concept ([3], p. 47). Claims of need have a special force in most cultures. Perhaps "P has a need for X" exercises a greater force than "P wants X". And if it does it is not because X in itself is considered valuable, but because X is necessary in order to reach a goal — which is considered valuable. (And in a longer need-chain containing subgoals, it is the final goal which is considered intrinsically valuable.) Claims of need get their force from the desirability or value of the goal. This explains why it is normally considered more important to meet the need for food (in order to survive) than the need for a luxurious car (in order, perhaps, to make a great impression on one's neighbours).

The goal is not specified in the definition noted above. A specified goal is necessary if the concept of need is to be useful in models of theories of various kinds. What then is the goal of need — that is, the goal that is to be reached when the need is completely satisfied? Most adherents of this tradition seem to answer the question by "avoidance of harm". J. Feinberg, for instance, believes that "P needs X" means "P will suffer harm if he doesn't have X" ([5], p. 111). However, "avoidance of harm" is an insufficient specification of the goal. It is now necessary to determine what is meant by "harm". There seem to be fundamentally different opinions on this issue. Let me present three of them.

-   P suffers harm if P does not live and develop in accordance with the good side of his or her nature.
-   P suffers harm if it is impossible for P to carry out the activities essential to his or her plan of life.
-   P suffers harm if P's interests are unanswered.

I will say more about these ideas in a moment. Let us just notice that it is when defining the goal of need that the concept of quality of life will become interesting. Quality of life may constitute the goal of need. We then have a fourth type of specification of "harm".

- P suffers harm if P's quality of life is low.

Let us now in turn look over these alternative specifications of "harm" — or, to put it another way, the goals of need.

DEVELOPMENT IN ACCORDANCE WITH HUMAN NATURE

H. J. McCloskey is of the opinion that it is detrimental to have to dispense with what we need. What will count as detrimental is not determined by, for instance, certain wants. In order to get an opinion on what is detrimental it is necessary to turn to human nature theories, says McCloskey. Needs relate to what it would be detrimental to us to lack, where the detrimental is explained by reference to our nature as human beings and specific persons. The detrimental should not, however, only be related to the way we are now but also to our natural development as individual persons, according to McCloskey ([10], p. 6).

It is possible that we as persons also have less pleasant sides. Shall a need be determined also by a development of the bad sides? No, it shall not, says McCloskey. Needs relate only to valuable existence and to valuable self-development, and we should only talk about need in relation to a good human nature and a good self-development (i.e., a development of something good): Talk of human needs and needs of particular persons involves references to natures, the perfection, development, non-impairment of which are good. Thus the goal of need, according to McCloskey, is *to exist and develop in accordance with our good nature*. This means that everything that is necessary for this goal is a human need, and things like food, love, social relations, meaningful activities, etc. are human needs only if they are necessary for this particular goal.

A FULFILLED PLAN OF LIFE

According to D. Miller, a person's identity can be established by the aims and activities which constitute his plan of life [11]. Without such an identity he can hardly be regarded as a person in the full sense, says Miller. A person's plan of life is constituted by the aims and activities which are central to that person's way of life, for instance what work he does, what social relationships he has, etc. Harm, according to Miller, is whatever interferes directly or indirectly with the activities essential to his plan of life.

Plans of life can take different forms. Miller gives the following examples: (a) carrying out a certain social role, (b) the pursuit of a social ideal, (c) a project such as cataloguing all the plants in South-East Asia, (d) the attempt to develop personal relationships of a particular type. Not every plan of life, however, will do, according to Miller. It must be intelligible to us. We must understand how, for the person who has it, that plan of life has significance and value; that is, we must be able to see how the person himself may value it, says Miller. Thus if confronted with a pyromaniac we are likely to say, not that he needs a plentiful supply of matches, access to barns, etc., but that he needs psychiatric help. According to this view, a person needs what it is necessary for him or her to have in order to fulfil his or her plan of life. The goal of need, according to Miller, is a *fulfilled plan of life* of the individual.

SATISFIED INTERESTS

There is a fundamental difference between the two goals above. A person's nature is in some sense more absolute than is a person's plan of life. The latter is chosen in a way that natures are not. However, the two goals need not be incompatible. It is for instance possible to choose as a plan of life to exist and develop in accordance with one's nature, or such existence and development could be a necessary condition for fulfilling the plan of life.[4]

The third alternative which focuses on the individual's interests seems to take an intermediate position. According to G. Thomson, a person P suffers harm when P's interests are not answered [17]. What then does Thomson mean by "interests"? Firstly, it is not identical with properties of human nature, but only dependent upon such properties.[5] Secondly, "interest" is not identical or equivalent to desires or wants. That a person's interests are satified does not necessarily mean that his or her desires are satisfied, and vice versa. However, interests are more or less directly related to the person's nature and desires.

Behind every desire there is a reason or motive. Some of our desires are instrumental, others are intrinsic. It is the reasons or motives behind the intrinsic desires that Thomson calls "interests". The following example may explain the relation between interest and desire. P has great ambitions and a strong desire to succeed in his professional life. Behind the desire there may be a striving for confirmation or appreciation. If so, achieving confirmation or appreciation is the reason behind the desire for success. P's interest in this case is the achievement of confirmation or appreciation. A person suffers harm, according to Thomson, when his or her interests are not satisfied. The person has a fundamental need for X when X is necessary in order to satisfy the person's interests. The goal of need according to this view is *satisfied interests*.

QUALITY OF LIFE

The concept of quality of life gives us a fourth way of defining the goal of need. The idea is reflected in the expression: "P suffers harm when P has a low degree of quality of life". This implies that P has a need for X when X is necessary in order to achieve or maintain a certain degree of quality of life. This idea is of little value, however, as long as the concept of quality of life is not analyzed. As is evident from the essays in this volume, there are different views on what we should refer to by the term 'quality of life'. However, one conceivable alternative is not possible in this case. We are here discussing a view of "need" where the concept is defined in terms of quality of life. We must then, in order to avoid circularity, refrain from defining quality of life in terms of need.

The concept of quality of life is related to the concept of a good life. It sounds like a contradiction to say, for instance, that P's quality of life is high but he lives a miserable life. A person's life is composed of several elements. Which of them are related to the good life or the quality of life? We seem to consider mainly two dimensions when assessing the good life. One concerns the material, psychological, and social conditions of the individual. Income, housing, employment, physical and mental health, and social relations are examples of such conditions. The other dimension concerns inner or mental states of the individual. Happiness and well-being are common terms for designating these states.

The term 'quality of life' seems in many cases to be reserved for states on the latter dimension, while 'welfare' seems to be a common term for the former dimension. S. Naess, for instance, use the term quality of life for designating states on the mental dimension: "These states are affective states, that is feelings, emotions, such as happiness — they are not cognitions, not judgements, such as satisfaction" [13]. In the book *Att ha, att älska, att vara (To Have, To Love, To Be)*, E. Allardt uses the term 'välfärd' (welfare) to designate material and social conditions [2].

A third dimension is presented in this volume. The term quality of life here refers not to inner or mental states (like feelings of pleasure), and not to material, psychological, or social conditions, but to an equilibrium between a cognitive state and a psychological state of the individual. The theory, put forward by Nordenfelt, does not involve an explicit definition of harm, but is nevertheless an interesting alternative for the goal of need. Roughly, a person has a high degree of quality of life when he or she is happy with his or her life. A person P is happy with life, according to the theory, when P's conditions in life are as P wants them to be. (See Nordenfelt's contribution to this volume.)

We have various goals or wants in life — some are important, others are less important, some are long-term, others are less long-term. We are happy with life when we perceive our wants as satisfied. Or more precisely, person P is happy with life if, and only if, P's conditions in life — as P sees them — are as P wants them to be. This aspect of happiness concerns the degree of equilibrium between, on the one hand, the person's wants concerning internal and external conditions, and on the other hand, P's apprehension of these conditions.

The basic thought in this theory is that a person has a high degree of quality of life when he or she is happy with life. The latter state, then, may constitute the goal of need. We then get: P has a need for X when (to have) X is necessary in order for P to live a happy life.[6]

Quality of life is a dimensional concept: a person could be more or less happy with life. This makes it possible to classify need with respect to degree of happiness. We might, for instance, with the help of the notion "minimal happiness" create two classes of need. P has a *fundamental* need for X when X is necessary in order to obtain or maintain a minimal degree of happiness, and P has a *marginal need* for X when X is necessary in order to obtain or maintain a degree of happiness above the minimal level. "Minimal happiness" might be related to, for instance, the satisfaction of high-priority wants.

We are here primarily dealing with the concept of need, and with a tradition according to which something is needed when it is necessary in order to achieve a certain goal. The significance of this goal has been pointed out — it is a determining and a justifying component of the concept. Basically we have to choose the goal of need, and I have presented four alternative specifications of it. A natural next step would be to analyze the similarities and differences between the alternatives. However, I shall here only comment on a few relations. (I have already commented on the relation between the human nature account and the plan of life account.)

The plan of life account of the goal of need (or "harm") implies that X is needed when X is necessary for fulfilling the plan of life. Fulfilling one's plan of life means realizing one's important wants. The plan of life account therefore has something in common with the quality of life account presented here. In both cases satisfied wants are involved. But the two accounts differ in several respects. First, the plan of life account excludes unintelligible wants — the quality of life account does not. Second, the wants are not related to happiness in the plan of life account, but they are in the quality of life account. However, that P's plan of life is fulfilled does not exclude the possibility that P has a high degree of quality of life, as when the wants which constitute P's plan of life are identical with P's high-priority wants. Third, the wants in the plan of life account concern the person's own life. The wants in the quality of life account may also concern situations outside the person's own life such as peace on earth, or no starving children in the world.

Satisfied interests constituted the third alternative definition of the goal of need. 'Interest' is here defined as the reason or motive behind a person's desire. The interest account and the quality of life account are therefore conceptually different. But the two accounts do not exclude each other — a person might simultaneously, by coincidence, have satisfied interests and a high degree of quality of life. In those circumstances the two accounts may generate identical sets of needs (or more correctly, objects of need).

I will now pass on to discuss an alternative way of looking upon quality of life, where the relation between the concepts of quality of life and need is the opposite to the one just presented.

QUALITY OF LIFE DEFINED IN TERMS OF NEED

Quality of life is here equivalent to need-satisfaction. The conceptual relation may be illustrated by the following statement: P's quality of life is high if, and only if, P's needs (or fundamental needs) are satisfied. However, the statement is inadequate as a definition of 'quality of life' because "need-satisfaction" is unanalysed — that is, the concept of need is not defined.

We could as before espouse the goal-related tradition when defining the concept of need, and consider needs (or objects of need) as necessities for realizing a certain goal.[7] But we could also follow the other tradition and consider need as a tension in the organism. Before I present the basic thought of the latter tradition I wish to make a few short reflections concerning the goal-related view of the concept of need.

I have argued that the goal of need (what is to be realized by satisfying the need) constitutes a significant component of the concept, and that the goal has to be defined if the concept is to be useful. I presented four alternative definitions. Three of them are applicable also in this case. The fourth alternative, where the goal of need is equivalent to quality of life, leads to circularity in this case (e.g., P's quality of life is high if P's needs are satisfied, and P's needs are satisfied if P's quality of life is high).

(A) The goal of need is to *exist and develop in accordance with the good side of human nature*. We then get the following structure: P's quality of life is high if P's needs are met, and P's needs are met if P exists and develops in accordance with the good side of human nature, or if P has at hand what is necessary for this.

(B) The goal of need is *a fulfilled plan of life*. Here we get the following structure: P's quality of life is high if P's needs are met, and P's needs are met if P's plan of life is fulfilled, or if P has at hand what is necessary for the fulfilment of the plan of life.

(C) The goal of need is *satisfied interests*. By "interest" is here meant the reason or motive behind a desire. In this case we get the structure: P's quality of life is high if P's needs are met, and P's needs are met if P's interests are satisfied, or if P has at hand what is necessary in order to satisfy the interests.

Defining "need" in accordance with the goal-related tradition could have a consequence worth noticing. It appears as superfluous to define something in terms of need. For instance, if concept B is defined in terms of need, and need in terms of Z, it would perhaps be much more simple to define concept B directly in terms of Z. Consider the following example: P's quality of life is high if P's needs are met, and P's needs are met when P's plan of life is fulfilled. Why not define 'quality of life' directly in terms of fulfilled plan of life — for instance, P's quality of life is high if P's plan of life is fulfilled.

In a study on quality of life by A. Aggernaes, the concept of quality of life is defined in terms of need [1]. Quality of life is equivalent to "satisfied fundamental needs". Fundamental needs are those needs which are common to all human beings in all hitherto studied cultures, and they are of such kind that human beings suffer if they are not satisfied. When is a need satisfied then? What conditions have to be fulfilled in order for a need to be considered as satisfied? Aggernaes is not completely explicit here. He seems to mean, according to my interpretation, that they are satisfied when "the human being doesn't suffer", when he "leads a good life", or when he "is happy and satisfied".

Perhaps it would have been more simple to say that P's subjective quality of life is high if P is happy and satisfied, and P's objective quality of life is high if P has the resources for becoming happy and satisfied (Aggernaes distinguishes between subjective and objective quality of life in this way).

Why then take a roundabout route via need when defining the concept of quality of life? One reason could be that it is desired to emphasize the distinction between the conditions of quality of life and the quality of life itself. What we need in order to achieve quality of life (material, psychological or social conditions) may constitute these conditions. And for practical purposes it might be easier to focus on the conditions — for instance, when institutions want to support individuals.

An alternative explanation might be that it has not been observed what the instrumental structure of the concept of need is — that we need things if they are necessary for a certain goal, and that the goal determines what objects we need. Therefore, perhaps, the tension view of need has been taken for granted.[8] Let us now take a look at this view of need.

NEED DEFINED AS A TENSION IN THE INDIVIDUAL

The 'term' need was introduced in academic psychology in connection with the development of theories of motivation. It there came to designate a state of disequilibrium. In the *International Encyclopedia of the Social Sciences* (1979) we read that "need is ... used by psychologists today to designate an internally or externally aroused, brainlocated force (often coupled with an accelerating emotion), subjectively experienced as an impulsion or felt necessity (a mild or intense urge) to act (immediately or later) so as to produce a certain specifiable terminal effect which is ordinarily expected to prove beneficial to the actor, and/or positively hedonic (less painful, more pleasurable) relative to the arousing situation."

P.W. Kurtz represents such a view of need [7]. He sees the living organism as a system in equilibrium. This system might be brought into imbalance through disturbances induced by external or internal stimuli. The lack of balance causes tensions. These states of tension-disequilibrium constitute needs, according to Kurtz. The organism is motivated to eliminate the lack of balance and thereby re-create a state of equilibrium. The process described here by Kurtz has much in common with what usually is called 'homeostasis'.

With this view of need we get the following relational structure between quality of life and need: P's quality of life is high if P's needs are satisfied. When are the needs satisfied, then? A reasonable answer is: "When there is equilibrium". A more common answer, perhaps, is: "When P has at hand what P needs" — that is, what P actually strives for. Some people believe that the assessment of needs can be done exclusivly by empirical means. When the individual is in need — that is, is in a state of disequilibrium — certain behaviours are triggered off towards certain objects. We can study these behaviours empirically, and gain knowledge about the object — that is, what is needed.[9]

We have the following characterization of quality of life if we apply this view of need-satisfaction: P's quality of life is high when P has obtained what P, because of inner tensions, actually has striven for. We may then ask if all the tensions are of equal worth. No one seems to answer such a question in the affirmative. Non-legitimate tensions are normally called false or artificial needs. According to Allardt [2], needs created through, for instance, advertising or whims of fashion are to be considered as artificial.[10]

An alternative way of assessing the needs of human beings is to turn to a need theory. A frequently employed theory in these circumstances is A. Maslow's psychological theory of need (despite its lack of empirical support; see [18]). Maslow postulates five basic needs, hierarchically ordered according to strength: physiological need, need for safety, need for love and belongingness, need for self-esteem, and need for self-actualization [9]. Suppose we let Maslow's theory of needs constitute the base for a definition of 'quality of life'. We then get, for instance: P's quality of life is high if P's basic needs are met, and P's basic needs are met when P has achieved security, belongingness, self-esteem, etc.

This way of looking upon need, as a tension in the organism, has been questioned. Some people think it is an inadequate use of the term need (e.g., [17]). Furthermore, to distinguish needs from preferences will be difficult when need is seen as a tension (for example, when a person buys cigarettes, does he need them, does he want them, or both?). In addition, it seems to be difficult to completely avoid a goal-related view of need. Maslow might be an example here. A few of Maslow's expressions could be interpreted in such a way that he believes that we need the food, security, belongingness, etc. in order to maintain our *health*. Thus we have a goal-related view where health is the goal of need.[11] Defining 'quality of life' by means of this alternative interpretation of Maslow can give the following structure: P's quality of life is high if P's needs are met, and P's needs are met if P is healthy, or if the conditions for P's health are fulfilled. Quality of life is here equivalent to health or the conditions for health.

We have now been discussing a type of relation where 'quality of life' is defined in terms of need or need-satisfaction. Two different ways of looking upon need might be involved in these circumstances: a) a goal-related view where a need can be seen as either a difference, or as a necessity for realizing a certain goal; and b) a view where need is seen as a tension in the organism. These two traditions of need might in certain circumstances coincide. A person who suffers from, for instance, shortage of body fluid and feels thirsty might be said to have both a goal-related need and a tension-need for water. In the case of a goal-related need, the goal might be either eliminated tension (P is no longer thirsty) or eliminated shortage of body fluid.

It is important, however, to notice that the two views of the concept of need are fundamentally different. To assess human needs in accordance with the goal-related view is to assess what things are necessary in order to realize a certain goal (the goal could be, for instance, a fulfilled plan of life). To assess human needs in accordance with the tension view is to register what people actually strive for. Assessing what people need (the object of need) in accordance with these views may therefore give totally different results.

We have so far only dealt with the logical relation between the concepts of need and quality of life. I conclude on the causal relation between need and quality of life.

SATISFIED NEEDS AS INDICATORS OF QUALITY OF LIFE

Suppose that satisfied needs lead to a high degree of quality of life. It is then possible to use need-satisfaction as indicator of quality of life. But it is not then a question of a logical relation between the concepts of need and quality of life. It is a causal relation exemplified by statement (1) at the beginning of the essay. (In [2] and [6] need-satisfaction is used as indicator of welfare.)

It is important when using need-satisfaction as indicator of quality of life that the concepts of need and quality of life are in harmony. Suppose, for instance, that we define 'quality of life' in terms of happiness and need in terms of fulfilled plan of life. Harmony means in this case that quality of life actually is a result of a fulfilled plan of life. Perhaps there is a certain risk of disharmony — in this case need-satisfaction in terms of fulfilled plan of life need not always lead to happiness. The risk is hardly less if we use a theory of need like Maslow's, for instance, where the conceptual foundations are more or less uncertain. Maybe the risk of conceptual disharmony is even greater if we consider need as a tension in the organism and at the same time define quality of life in terms of happiness. It is true that tensions trigger off behaviors, but it is uncertain whether they automatically direct the individual's behavior towards things which are conducive to happiness. Need-satisfaction will not be a valid indicator of quality of life unless the concepts of need and quality of life are properly related to each other.

SUMMING UP

Need and quality of life can be conceptually related in two principal ways: 'need' may be defined in terms of quality of life, and 'quality of life' may be defined in terms of need. According to the goal-related view of need, something is needed if it is necessary in order to realize a certain goal. The goal has a central place in the concept, and four different alternative ways of defining the goal have been presented. Quality of life is one of these. Fulfilled needs are necessary for quality of life according to this way of defining the goal of need. However, 'quality of life' in turn has to be defined, if we want an adequte definition of need. I have roughly presented the basic elements of a theory where quality of life is equivalent to happiness (as an emotion).

In the second type of relation 'quality of life' is *defined* in terms of need — for instance, "P's quality of life is high" means that P's fundamental needs are satisfied. Two different views of when a need is satisfied have been presented. These views are based on different opinions on the nature of need. A need is satisfied, according to the goal-related tradition, when the goal is realized or when the individual has at hand what is necessary for realizing the goal. Suppose that the goal is a fulfilled plan of life. P's quality of life is high, then, if the plan of life is fulfilled. From this we might draw the conclusion that it is superfluous to define "quality of life" in terms of need.

Need may also be considered as a tension or disequilibrium in the organism. The tension triggers off behaviors towards certain objects. Getting hold of these objects is expected to lead to reduction or elimination of the tension. Needs are satisfied and quality of life achieved when the tension is eliminated. We can determine what people need, according to this view, by studying people's behavior. Another way of determining what people need would be to turn to a psychological theory of need. In this case it is important to know the conceptual foundations of the theory. The foundations of, for instance, the often-used theory of Maslow's seem to be rather uncertain. There are therefore reasons for being careful when using this theory for characterizing quality of life. Furthermore, the demand for validy requires that the relation between the concepts of need and quality of life is harmonious when using need-satisfacton as indicator of quality of life.

NOTES

[1] This is a revised version of an article published in the Norwegian journal Norsk Filosofisk Tidsskrift 1991.

[2] The relation between the concepts of need and health is analysed in [8] and [14]. The relation between the concepts of health and quality of life is dealt with in [15].

[3] See ([8], pp. 45-47). According to T. Moum, focusing too much on the object of need might lead to a working out of long lists of needs. "Very often one ends up with taxonomies that are extremely detailed and completely unresearchable" ([12], p. 8).

[4] But the two states can also be in conflict, and if one sides with McCloskey it is possible to argue like Braybrooke: "People have a need for exercise regardless of what they wish, prevent, want otherwise, or choose. They have the need even if they do not much care to live or be healthy" ([4], p. 32).

[5] To exist and develop in accordance with the good side of human nature is probably not equivalent to having satisfied interests. But according to A. Maslow's view of human nature a person is motivated under certain circumstances to strive for self-realisation. If this is correct, and "self-realisation" is equivalent to "exist and develop in accordance with one's nature", the goals would in practice, for one and the same person, in similar circumstances, generate identical sets of needs.

[6] The statement shows that the term "quality of life" might be superfluous when defining the concept of need. "Need" might be defined directly in the terms used to define "quality of life".

[7] The idea of an equivalence between need-satisfaction and quality of life is discussed by Moum [12]. Need is here related to happiness. Needs are things which contribute to maximal happiness. Moum's treatment of need is an example of an analysis which mainly focuses on the object of need.

[8] Maybe this applies to Aggernaes' case [1]. He declares the following view about needs: "By 'needs' are understood tendencies of action towards specified goal-situations, which could be of the categories to achieve, maintain, or avoid specific states, courses of events or situations.... When needs come into existence or increase in force, we will typically experience wishes for, yearning for, craving for, or desire for the thing which is the goal-situation of the need" (p. 58). (My translation)

[9] In a report on welfare research and social policy we find the following opinion: To say that someone has a need for X is to say that one has empirically established that human beings during a certain period, with a certain regularity, acquire X, and among those people who haven't acquired X during this period we notice an ever-increasing motivation to provide X ([16], p. 186). Also, E. Allardt believes that it is possible to assess what we need through empirical observations of the individual's behaviour ([2], p. 19).

[10] Aggernaes seems to be of another opinion. He doesn't speak of false needs — a person's quality of life is high irrespective of what kind of tensions are involved. A person may, for instance, increase his quality of life by bullying someone. Aggernaes distinguishes, however, between two kinds of quality of life. One that should be noted by, or attract interest from, the institutions of society, and another that should not [1].

[11] The following quotation illustrates Maslow's goal-related view of needs where the goal is some kind of state of health: Basic needs are human goods that "are not only wanted and desired by all human beings, but also needed in the sense that they are necessary to avoid illness and psychopathology" ([9], p. xiii).

BIBLIOGRAPHY

[1]   Aggernaes, A.: 1989, *Livskvalitet*, Foreningen af danske laegestuderendes forlag, Copenhagen.

[2]   Allardt, E.: 1975, *Att ha, att älska, att vara: Om välfärd i Norden*, Argos, Lund.

[3]   Barry, B.: 1965, *Political Argument*, Routledge & Kegan Paul, London.

[4]   Braybrooke, D.: 1987, *Meeting Needs*, Princeton University Press, New Jersey.

[5]   Feinberg, J.: 1973, *Social Philosophy*, Prentice-Hall, Englewood Cliffs, New Jersey.

[6]   Köhler, L. (ed.): 1990, *Barn och barnfamiljer i Norden. En studie av välfärd, hälsa och livskvalitet*, NHV-Rapport 1990:1, Lund.

[7]   Kurtz, P.W.: 1958, 'Need Reduction and Normal Value', *Journal of Philosophy*, 55, 555-567.

[8]   Liss, P-E.: 1993, *Health Care Need. Meaning and Measurement*, Avebury, Aldershot, U.K.

[9]   Maslow, A.H.: 1970, *Motivation and Personality*, 2nd ed., Harper & Row, New York.

[10]   McCloskey, H.J.: 1976, 'Human Needs, Rights and Political Values', *American Philosophical Quarterly*, *13*, 1-11.

[11]   Miller, D.: 1976, *Social Justice*, Clarendon Press, Oxford.

[12]   Moum, T.: 1990, *Needs, Rights and Indicators in Quality of Life Research*, Institute for Social Research, Oslo.

[13]   Naess, S.: 1989, 'The Concept of Quality of Life', in Björk, S. and Vang, J. (eds.), *Assessing Quality of Life*, Health Service Studies 1, Linköping Collaborating Centre, Linköping.

[14]   Nordenfelt, L.: 1987, *On the Nature of Health*, D. Reidel Publishing Company, Dordrecht.

[15]   Nordenfelt, L.: 1991, *Livskvalitet och hälsa: teori och kritik*, Almqvist & Wiksell, Stockholm.

[16]   Ohlström, B.: 1975, 'Om mänskliga behov och sociala indikatorer - en referensram', in *Velferdsforskning og sosialpolitikk*, Rapport fra det 11. nordiske sosialpolitiske forskerseminar, Ringen, A. (ed.), INAS rapport 1975:1, Oslo.

[17]   Thomson, G.: 1987, *Needs*, Routledge & Kegan Paul, London.

[18]   Wahba, M.A., and Bridwell, L.G.: 1976, 'Maslow Reconsidered: A Review of Research on the Need Hierarchy Theory', *Organizational Behavior and Human Performance*, *15*, 212-240.

NEEDS, RIGHTS AND RESOURCES IN QUALITY OF LIFE RESEARCH[1]

In what follows I plan to come to grips with some terminological and philosophical confusion that seems to underlie a number of controversies between what is often viewed as rather antagonistic approaches in the study of human welfare. The main argument will be that a utilitarian, "subjectivist" quality-of-life approach is not necessarily inimical to a conception of welfare research oriented toward "needs", "resources" or "rights". I shall discuss in some detail a conceptual structure in which quality of life *qua* subjective well-being (happiness/utility) is presently being used as an empirical referent and as the "hard core" normative takeoff point for delineating such elusive concepts as "needs" and "rights". Clarifying the relationship between these concepts and approaches may eliminate some terminological confusion as well as provide some indication of areas of substantive research that deserves special attention in the years to come.

Much of the impetus behind my analysis is a conviction that from a moral point of view, Anglo-American "quality of life" research as we have encountered it over the last two decades must be seen as a basically utilitarian philosophical enterprise. It does not take much imagination to appreciate the very close link between the *summum bonum* of classical utilitarianism and the "subjective well-being" focused on by empirical quality-of-life researchers. However, the utilitarian underpinnings of quality-of-life research have not been sufficiently reflected on, and, paradoxically, it seems as if *precisely because of their subjective nature*, welfare differentials as assessed by quality-of-life researchers are not often taken quite seriously in a (normative) policy context, or in "level-of-living" research[2].

One of the most common objections invoked against equating "utility" with "welfare" is the claim that when individuals are dissatisfied or unhappy this may simply be the result of thwarted *desires* or *wants* to which conventional wisdom or our moral "intuitions" are not willing to grant priority [15] [28]. The idea that human desires are amorphous, unreliable and bottomless has, of course, a long-standing tradition in Western philosophy and sociology from Rousseau and Durkheim onwards (see e.g., [33]). Nevertheless, I shall try to show that empirical research on human welfare

79

L. Nordenfelt (ed.), Concepts and Measurement of Quality of Life in Health Care, 79–93.

(not least in affluent, Western societies) does require some sort of minimal conceptual framework and operational procedures which allow for interpersonal comparisons of utility. Otherwise one very easily runs the risk of one's results becoming normatively and politically irrelevant. In other words, I am not convinced that needs, rights or resources are alternative core ideas which are capable of by-passing some of the difficulties inherent in a hard-nosed subjectivist, utilitarian formulation in welfare research.

One of the reasons for addressing these questions is that the so-called Scandinavian approach to welfare research very consistently has opted for the *resources* available to individuals as the best basis for making interpersonal comparisons of welfare [7]. Part of the rationale behind this choice obviously has been that resources are more relevant to the political process [11], but in addition there has been a strong preference for "descriptive" or objective methods of measurement, to which resources presumably are more amenable. Outside of Scandinavia *needs* (or basic, fundamental, "real" needs) and the extent to which they are satisfied has been a very popular choice as a core idea when defining welfare at the level of the individual [12] [29]. Of course, resources (often in conjunction with "conditions") most commonly are viewed as being more or less synonymous with "need satisfiers" so that there does not seem to be too great a discrepancy between a needs approach and a resource approach. Since needs appear to be a primary category relative to resources (resources attract interest primarily because they are need satisfiers), I shall concentrate my efforts on explicating the link between *needs* and happiness/utility. In agreement with Liss [13], I think it is important that this relationship be reflected upon, and that one is clear in one's mind whether the link between need and happiness/utility is conceived as a *logical* or an *empirical* one.

Defining welfare as the satisfaction of (basic, real, fundamental, universal, legitimate) needs rather than as happiness or utility apparently allows us to steer clear of the complications that arise when "desires", "wants", "relative deprivation", etc. make a claim to being relevant to welfare (as they would if happiness were the core concept). In a similar fashion, some authors try to avoid considerations of happiness and the "hazy relativity of desires which ... reach out for the infinite" by stressing the primacy of (individual) *rights* in a normative policy context [9]. Hersch warns us that we should not romanticize the issue of human rights by identifying rights with happiness for all, or with the permanent establishment of universal peace. The *absolute* roots of human rights would be destroyed if we commit such a fallacy. Pursuing such a line of reasoning one could of course question the general idea that human welfare should be couched in subjective or utilitarian terms (see, e.g., [23]). Thus Wilcox claims that the notion that welfare can be subjectively defined would render most philosophical discourse on ethical issues irrelevant, leaving "... not much more than a combination of ethical relativism and metaethical emotivism" ([31], p. 8).[3]

My impression is that attempts to discard utilitarianism in welfare research (see, e.g., [25]) nevertheless seem to rest on implicit assumptions about the preconditions of individual utilities (see [24]). Depending somewhat on one's level of ambition (do quality-of-life considerations extend to all living systems or even to the entire biosphere?), a utilitarian approach at least seems able to accommodate important *aspects* of the normative reasoning implicit in a rights perspective (e.g., through various forms of "rule utilitarianism" and the like (see [14] [21]. In other words, if one adheres to what Elster [6] refers to as *ethical* individualism ("in the final analysis only individuals are morally relevant"), one simply cannot *discard* subjectivist, utilitarian considerations in a normative context, even though it is quite possible to provide examples to show how intractable some interpersonal comparisons of utility may be [5].

WELFARE, NEEDS, RIGHTS AND SUBJECTIVE WELL-BEING

Are there indeed such entities as "fundamental" or "basic" needs and rights which should remain conceptually distinct from subjective utility or happiness? The need concept, time and again, has emerged as an almost irresistable lure to quality-of-life researchers. There are literally dozens of theorists who have proposed to equate life quality with need satisfaction (see [29] [12] [27] [1]). There is usually a shift in emphasis in the direction of working out exhaustive lists of "basic needs", personality models, etc. Frequently one ends up with taxonomies that are extremely detailed but completely unresearchable.

Or, in order to obtain conceptual clarity, one proposes *empirical* need models which are blatantly unrealistic (e.g., stating that needs are satiable, hierarchically arranged, and of the same magnitude across all individuals [17]). Other need models are symbiotically tied to one particular brand of personality theory a very popular candidate is Maslow's [30], or simply void of substantive content e.g., because the definition of needs is deferred to an arbitrarily chosen group of experts [3].

Much of this conceptual confusion seems to arise not only because of the lack of terminological clarity and precision, but because "needs" are

constantly invoked in *substantively quite distinct fields or contexts of
enquiry*: Partly (1) the need concept is utilized within various psychological
theories of personality organization as well as in everyday language for purposes of *motive imputation*; partly (2) it is used *descriptively* to account for
various responses of the organism to diverse deficiencies in the environment
(when people's needs are not satisfied they die, lose weight, become
mentally ill, etc.); partly (3) it operates as a normative category (people
ought to have their needs satisfied). There is no logically necessary link
between these three uses, and several problems may arise if one fails to
recognize these distinctions. In particular I think the following tack is very
often taken: The rationale for choosing the need concept in the first place
springs from one's underlying philosophical convictions (presumably
utilitarian), but because of the ambiguity of the need concept, one struggles
onwards in order to attain analytical neatness and closure with respect to the
motivational and descriptive uses of the concept. More often than not such
closure will be premature, and one runs the risk of letting an overly
specified need concept detract from one's ability to tackle the problems
discussed above.

   In contract to such an approach, I propose we adopt a strictly normative
(utilitarian), "formal" definition of needs and *leave it at that*. Instead of
trying to work out elaborate taxonomies of needs according to their
"basicness", I shall suggest several rules ("guidelines", cf. [26]) which may
help quality-of-life researchers select variables and models, as well as chart
need structures, thus aiding policy makers in assigning priorities to needs.

Consider the following definition of needs:

*If it can be shown that a person P obtains more happiness or utility from a
given (possible) state of affairs A than from another possible state of affairs
B, then P may be said to have a need for A (relative to B).*

At first glance this may have the appearance of a simple preference relation,
but it is not. Of course, I am not making the assumption that P actually
wants A or that he or she attaches any expected utility to its realization.

   'State of affairs' is intended to be totally comprehensive, i.e., to cover
all observable and unobservable variables, including intrapsychic properties
of a need subject. If lowering the level of blood sugar for a given individual
positively affects his or her experience of utility, then *ipso facto* this
individual has a need for less blood sugar. Also, the term 'state of affairs' is

quite unspecified in its temporal aspects, hence it should cover both recurring, cyclical needs (such as the needs for nutrition and energy) as well as needs pertaining to stable states (housing, systems of transportation, etc.). This would correspond to Drewnoski's [3] distinction between "flow" and "stock" of welfare. (When talking about "a" given need, one necessarily operates with clusters or classes of states of affairs, depending on one's level of abstraction; the "need" for nutrition, etc.).

Needs as so defined may appear a hopelessly broad category, covering also feeling states and motivational categories such as "desires" and "wants", precisely those entities to which the "objectivist" approach to welfare is deeply sceptical. However, as I shall try to establish in the following, one of the main advantages of the proposed definition is precisely that it does allow one to make a valid normative distinction between needs and desires within a framework of need rules or guidelines.

There may be a further objection that the proposed definition, because of its comprehensiveness, does not take us one iota beyond the Greatest Happiness Principle. The operative ethical doctrine in this case simply would be "satisfy needs!" — which appears entirely parallel to the utilitarian "increase happiness!". In the main this is correct, although it is important to notice that "need" is seen as a *relational* concept: A need is a relation between an historically unique individual and a constellation of observable, environmental variables.

As already noted, however, the most obvious objection to the proposed definition is that the number of needs in principle becomes infinite, and we are left completely in the dark as to which needs ought to be satisfied and in what order this should happen. This immediately brings us to the question of specifying rules or guidelines for the weighing of needs so that the concept of need may play a constructive role in quality-of-life research. Specifying such rules may have the appearance of a relatively straightforward, logical exercise, in the process certain weaknesses in the established practice of quality of life research will be uncovered, weaknesses which might otherwise have passed unnoticed. In that case the scheme I shall propose will have proved its usefulness.

The list of rules or guidelines put forward must necessarily be tentative and incomplete at this point. I shall comment upon the following rules/guidelines: *potency, compatibility, growth, interaction, cohabitation,* and *viability.* Following Sundby [26] I shall not try to work out meta-rules regulating possible conflicts between individual rules. Indeed, one of

Sundby's main points is that such attempts in a philosophical (moral) context appear rather futile.

(1) *The rule of potency* states that needs should be given priority according to the short-term *amount* of life quality (happiness) to which they give rise. to. Needs which if left unsatisfied threaten to annihilate the individual in his or her physical existence or psychological integrity obviously qualify as potent. Usually they are referred to as "basic". Needs which show relatively small variations between individuals — presumably because they have a strong biological or genetic component — are usually called "universal". In general one would expect universal needs also to be very potent (basic), but this is not necessarily the case. Thus the need for variety and change, for travelling, seeing new places, etc., is indeed very widespread (universal), but if thwarted it probably does not cause major psychological damage in the need subject.

Potency enters as a crucial dimension in the definition of human *rights*. If society knows how to satisfy a potent need and if the resources required for its satisfaction are available, the procurement of such a need would tend to be viewed as a basic right. It is interesting to note that all three conditions (i.e., potency, knowledge and resources) are relevant for a given need satisfaction to qualify as a right. For example, there are certain potent needs (such as the need for self-respect and human dignity) for which we *probably* have the necessary resources, but where the ways and means of satisfaction are so complex (i.e., unknown) that we baulk at proclaiming the satisfaction of such needs as a basic right. Conversely, the satisfaction of needs which do not yield particularly high amounts of pleasure or happiness may be viewed as a right if the satisfiers (means) are well known and of small cost, e.g., the treatment of minor ailments, a broken nail, etc. If this analysis is correct, we may say that quality-of-life research may be instrumental in the definition of human rights, either by uncovering the mechanism by which needs are satisfied or by pinpointing particularly potent needs. Figure I illustrates this line of thought. The area north-east of the indifference curves covers those need satisfactions that qualify as rights. If there is a general increase in the resources available for need satisfaction, the field of rights would be extended, e.g., move from "B" to "A".

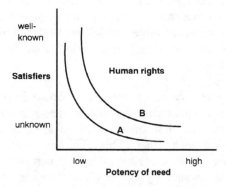

FIGURE I. Human rights defined as a function of need potency and level of knowledge
concerning its satisfaction

The potency of needs has of course been a major concern among life quality
researchers. A recurrent theme in life quality surveys has to do with the
independent contribution of various "life domains" to overall life quality,
clearly a property relevant to the understanding of the potency of the related
needs. However, such an approach does have its limitations. In addition to
possible biases in researchers' selection of domains or needs, the indeterminate relationship between the need subject's actual behaviour and the range
of possible need satisfactions must be considered. People may be *unaware of
their needs* because of lacking information about action alternatives. Or their
own life history and social background may have prevented them from
gaining insight into their own potentialities. The musically endowed individual who has never been exposed to music or tried to play an instrument may
unwittingly forfeit profound pleasures. This would be a case of a highly
potent need (for exposure to music and musical training) which could remain
dormant indefinitely. Such a possibility is not far-fetched. Autobiographies
abound in accounts of how individuals experience sudden and coincidental
revelations about the nature of their most potent needs — a vocational
calling or a hobby suffusing their lives with meaning and direction.

It is of course very limited what quality-of-life researchers can do with
respect to such problems when they have to rely on cross-sectional data.
However, simply charting the impact of differentials in the diversity of
experiential histories on present need structures may shed some light upon

the significance of different *life styles* or ways of life in laying the groundwork for potent needs. Related to this question is that of *level of aspiration*. Is there a clear-cut *linear* (inverse) relationship between aspirations and need potency? The exact shape of this function may determine how and where one should look for possible *optima* in levels of aspiration and expectations within a given society. This is indeed a challenging task for quality-of-life researchers.

(2) *The rule of compatibility* states that needs should be given priority according to the extent to which they are able to enter into *compatible sets*, i.e., whether needs may be simultaneously (or at least serially) satisfied without interfering with each other. This guideline refers primarily to intrapsychic processes. Do we have needs which — if they are accommodated — may seriously hamper or block altogether the satisfaction of other needs? There seems to be a standing presumption, at least from Rousseau onwards, that our needs (or our "real", "natural" needs) are bound to work harmoniously together. A prominent modern proponent of such a view is Abraham Maslow [16], whose humanistic psychology clearly presupposes that there is a natural, biologically given groundplan from which our needs evolve in perfect order and harmony (given benign circumstances, of course). The world view of evolutionary biological theory seems to differ radically from this position: Rather than being designed to maximize happiness our need structures must be seen as the result of a *best compromise* between a number of competing environmental, social and organismic mandates favoring inclusive genetic fitness. According to such a view one might hypothesize that for human beings in their natural condition (i.e., as hunters and gatherers), natural selection would tend to produce need structures which, given the restrictions of this mode of existence, yield the highest possible amount of pleasure. This is so because of the greater genetic fitness of individuals who find pleasure in doing what they must do (phylogenetically speaking). But this is of course a far cry from claiming that our "natural" needs do not tend to cancel each other out with respect to the happiness they produce. In particular, under highly artificial societal circumstances one would suspect that the net happiness result would be better if certain needs were suppressed altogether. Needs for aggressive behavior, dominance, and power if allowed to find their outlet may be detrimental to the satisfaction of the needs for belongingness and love for one and the same individual. This does not mean that dominance needs do

not belong to "human nature", it simply means that a given need does not *by its very existence* have a (moral) claim to be satisfied [2].

In quality-of-life research problems of compatibility arise, for example, in connection with the question of the *additivity* of life domains. Empirical work so far seems to suggest that additive models work well, but we should keep the possibility open that domains must be thought of as compatible sets, both in terms of their content and emphasis. Way of life studies may be of great help in this respect. Perhaps life styles should be thought of as a limited number of need constellations (compatible sets), which people choose by their own design, and not (as is the general rule) as culturally determined roles?

(3) *The rule of growth* states that needs should be given priority according to the extent to which they give rise to sustainable (long term) pleasures. Alcoholism and drug addiction may be condemned on utilitarian grounds because they violate the rule of growth. Intoxication may give momentary relief and pleasure, but such happiness cannot be sustained on a long-term basis. Thus put, one might say that the rule of growth is simply a corollary of the rule of potency since we are really talking about life time potency in this case.

There is, however, another twist to this rule: It also covers needs that are *necessary steps* in developmental sequences whose net result in terms of happiness may be shown to be greater than the repeated satisfaction of stable (unchanging) needs. In modern societies it seems as if social class differentials in life quality increasingly will be a function of limited possibilities for satisfying growth needs. The less educated in highly differentiated societies seem particularly prone to engaging in projects where action very soon is bogged down by routinization and a lack of prospects. From the point of view of maximizing life quality the distribution of material and non-material rewards over the life cycle seems highly sub-optimal in the case of unskilled labour. An important task for quality-of-life research must be to investigate life cycle inequities by trying to arrive at realistic estimates of the mandates of the division of labour ("functional necessities") as weighted against cultural inertia and prejudice.

(4) *The rule of interaction* states that the needs of an individual should be given priority if they are instrumental in satisfying the needs of those that the individual directly interacts with. One might also call this a rule of altruism. In a sense this rule is simply a generalization of the rule of

compatibility at the level of interpersonal interaction. Obviously, if an individual finds pleasure in catering for the needs of others, overall happiness stands to gain from this. However, it is somewhat misleading to think of the rule of interaction simply as an altruistic criterion. For example, the smooth functioning of an economic system seems to depend upon a peculiar *mixture* of egoistic and altruistic motivations (maximizing economic utility while abiding by the rules). Supposedly there may be negative spin-off effects also from the satisfaction of needs which are neutral along the altruism-egoism scale. In the main, however, the needs covered by the rule of interaction would be the "sadistic" motives (competitiveness, invidious comparison etc.), e.g., those which spill over from the greater, surrounding society into the spheres of family and friendship relations.

The rule of interaction also should cover needs whose satisfaction *presupposes* zero-sum conflicts, such as attaining a job in the upper echelons of the status hierarchies, gaining exclusive access to scenic beauty (e.g., through the acquisition of second homes) and other goods in absolute scarcity [10] [8]. In general, quality-of-life researchers have been rather oblivious to the dis-economies arising from the pursuit of such motives. Unfortunately, this practice probably is not simply a sin of omission, it seems to spring rather directly from the atomistic approach of the survey. If a respondent reports that he derives great pleasures from his high status job, no ledger is being kept where the detrimental effects on his fellow men from status inequalities and invidious comparisons are explicitly taken into account. One might retort that such effects would turn up as reports of lower levels of satisfaction among the less fortunate, but such an observation would not be integrated into a systemic perspective (where the interdependency between the two facts is captured), and it would not take into account the subtle processes of compensation, lowering of levels of aspiration, etc. among those who lose out in the competition [19]. As I have commented upon above, the net result of such processes is not all obvious.

In all probability, quality-of-life projects aiming to supply empirical material relevant to the rule of interaction cannot work simply with broad samples - communities, factories and smaller social units would be more appropriate.

(5) *The rule of cohabitation* states that needs should be given priority if they may be satisfied simultaneously for larger numbers of individuals. This rule may be seen as a simple generalization of the rule of potency. Population growth has made us acutely aware of the disasterous results which may

follow in the aggregate from practices which are quite benign on a small scale. Erosion, pollution and crowding are but a few of the innumerable examples. In many cases it may be difficult to distinguish between the pure effect of great numbers and the impact of system structure (i.e., processes more akin to those discussed in the previous section). Also, as quality-of-life researchers we should probably not aim at too high a level of sophistication in this area. Counterfinality in larger social systems may be uncovered without the aid of reports about individual welfare, and the remedies for such imperfections usually would have to be sought at the level of policy making. Nevertheless, life quality researchers may supply valuable predictions about the probable happiness results of the required changes in life styles and about the ways and means for bringing about such changes.

(6) *The rule of viability* states that needs should be given priority if they do not negatively affect the need satisfactions of future generations. This rule speaks for itself. It simply extends the Greatest Happiness Principle also in time. The tasks that this rule sets for quality-of-life researchers would be very similar to those posed by the rule of cohabitation. The implications generally would concern the relationship between life style preferences and social system functioning. Presently our polities seem ill equipped to cope in a satisfactory manner with the long-term consequences of current social practices. There are few built-in mechanism for preventing politicians and other groups who are wielding power from acting rather narrow-mindedly on behalf of their own constituencies and with very limited time perspectives. Also, it is extremely difficult to assess realistically the possibilities and challenges even of the next few decades. When our methods of social forecasting have been perfected, quality-of-life researchers could play a more active role in these matters.

CONCLUSIONS AND DISCUSSION

The present section has taken as its point of departure the contention that quality-of-life researchers rely on a clear-cut emotive definition of welfare, in practice setting out to measure the *summum bonum* of classical utilitarianism.

Working on the assumption that utilitarianism and quality-of-life research do in fact operate with common premises, one may also take it for granted that they have unresolved problems in common. The meaning of happiness, the relationship between choice (behavior) and utility, and the relationship between happiness, rights, and justice are pertinent issues.

Such common premises and problems seem to call for a conceptual framework which allows the quality-of-life researchers to talk about the empirical preconditions for maximizing welfare and relationship between individual welfare and public policy in a coherent manner. It is suggested that the *need* concept represents a fruitful core idea in this connection. An extremely broad, rather formal, definition of needs is put forward, and it is claimed that many of the problems common in life quality research and utilitarianism may be accommodated by a series of rules or guidelines intended to specify the conditions for need priority. In a somewhat simplified manner one might say that these rules are differentially located in a Cartesian space, where the axes are time and space, i.e., these rules take into account *future* need satisfactions and the needs of *other* individuals. (Yet the rules are of course designed to generate priorities for a given individual's needs.) The location of the rules are indicated in Figure II.

What are the advantages and drawbacks of such a scheme? Space does not allow me to discuss in detail the pros and cons of the proposed definition of needs. Let me briefly mention the following points:

(1) The proposed scheme allows us to stay within the same (utilitarian) idiom even when discussing quite diverse "values" or desiderata. Quite often a problem posed as competing claims between intrinsically incommensurable values should be thought of as a very complex calculus, with an underlying common denominator. This does not mean that a given problem is more easily solved, but it may imply that one is not unduly confused or caught in a quagmire of value theoretical or philosophical complications. Thus, some rather intractable problems arising within the "needs/resources" approach (which should be considered "critical points" for a given resource, etc. [4]), may be empirically elucidated by relating resources and conditions directly to subjectively assessed quality of life.

(2) The need concept is in good accord with everyday language and the common sense understanding of life quality. As Nielsen [22] puts it: "If someone needs something there is a standing presumption that he ought to have it." Even though the proposed rules rather consistently negate such a supposition, it is still very useful as a point of departure. Moreover, the idea

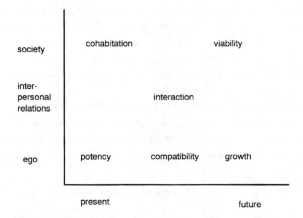

FIGURE II. Location of rules for need priority in time and "space"

that needs must be balanced against each other is not alien to the public mind. In general, adopting a terminology which accords well with common usage is well advised if one wants to promulgate communication with policy makers and others outside the scientific community.

(3) The proposed scheme is *empirically open*. Above all this means that the understanding of needs is not tied to any particular brand of psychological theory, nor does it presuppose that needs are stable and consciously appreciated by the need subject. In accordance with a Marxian or a psychodynamic view, one may expect that needs to a large extent are socially determined, but the possibility is not ruled out that empirical research will uncover a stable, genetically based, hard core in apparently "acquired" needs which demand satisfaction regardless of socialization practices or socio-cultural context.

Others are probably better suited for commenting upon the weaknesses of the proposed scheme. Let me point out only that the present approach obviously has a very monistic or "teutonic" slant to it, and that one may easily grow overly enthusiastic about one particular line of thought. Also, my perspective is quite clearly very *individualistic* in its basic emphasis. As I see it, this is no major draw-back in quality-of-life research, but one should have no illusions about the possibilities for applying the proposed framework mechanically to other social science disciplines.

NOTES

1 I am grateful for comments, criticism, and encouragement regarding earlier versions of this essay from Fredrik Engelstad, Arne Mastekaasa, Alex Michalos, Lester Milbrath, Siri Næss, Arne Næss, J.P. Roos, and Elèmer Hankiss, none of whom may be held responsible for my conclusions. This chapter is a revised and expanded version of Moum [20].

2 Part of the resistance against quality-of-life research also stems from scepticism regarding the validity of self-reported subjective well-being.

3 It is beyond the scope of the present essay to pursue the idea that welfare or quality of life should be based on theories of justice, i.e., on some sort of Kantian *Pflichtetik*, Sartrean authenticity, freedom of choice, self-actualization, eudaemonism or on a more or less self-contained theory of rights (see also McGill [18]).

BIBLIOGRAPHY

[1]   Aggernaes, A.: 1992, 'Generel og behovsspecificeret lykke og tilfredshet: livskvalitet som slutmålet i helsearbejde', *Socialmedicinsk tidskrift, 69*, 14-19.

[2]   Campbell, D.T.: 1979, 'Comments on the Sociobiology of Ethics and Moralizing', *Behavioral Science, 24*, 37-46

[3]   Drewnoski, J.: 1974, *On Measuring and Planning the Quality of Life*, Mouton, The Hague.

[4]   Elstad, J.I.: 1983, Temaer omkring en generell sosial rapport (*On a general social report*), (Report 83:6), Institute of Applied Social Research, Oslo.

[5]   Elster, J., and Roemer, J.E.: 1991, Introduction. In Elster, J. & Roemer, J.E., (eds.), *Interpersonal Comparisons of Well-being*, Cambridge University Press, Cambridge, Massachusetts, pp. 1-16.

[6]   Elster, J.: 1986, *An Introduction to Karl Marx*, University Press, Cambridge.

[7]   Erikson, R., and Åberg, R.: 1987, *Welfare in Transition. A Survey of Living Conditions in Sweden 1968-1981*, Clarendon Press, London.

[8]   Frank, R.H.: 1985, *Choosing the Right Pond*, Oxford University Press, Oxford.

[9]   Hersch, J.: 1986, 'Human rights in Western thought: Conflicting dimensions', in *Philosophical Foundations of Human Rights*, Unesco, Paris, pp. 131-148.

[10]   Hirsch, F.: 1976, *Social Limits to Growth*, Harvard University Press, Boston.

[11]   Johanson, S.: 1979, *Mot et teori för social rapportering*, (*Towards a Theory for Social Reporting*), Swedish Institute for Social Research, Stockholm.

[12]   Lederer, K.: 1980, *Human Needs*, Oelgeschlager, Gunn & Hain, Cambridge, Massachusetts.

[13]   Liss, P.E.: 1991, 'Behov og livskvalitet', *Norsk filosofisk tidsskrift, 26*, 215-231.

[14]   Lyons, D.: 1967, *Forms and Limits of Utilitarianism*, Clarendon Press, Oxford.

[15]   Mallmann, C.A., and Marcus, S.: 1980, 'Logical clarifications in the study of needs', in K. Lederer (ed.), *Human Needs*, Oelgeschlager, Gunn & Hain, Cambridge, Massachusetts, pp. 163-185.

[16]   Maslow, A.H.: 1970, *Motivation and Personality*, Harper and Row, New York.

[17]   McCall, S.: 1975-76, 'Quality of Life', *Social Indicators Research, 2*, 229-248.

[18]   McGill, V.J.: 1967, *The Idea of Happiness*, Praeger, New York.

[19]   Moum, T.: 1983, *Resignation and quality of life*. Paper presented at the 1983 World Conference on Systems, Caracas.

[20]   Moum, T.: 1990, *Needs, rights and indicators in quality of life research*, Working paper AN 90:5, Institute for Social Research, Oslo.

[21]   Næss, S.: 1976, 'Livskvalitet som målsetting for velferdsforskningen', *Tidsskrift for Samfunnsforskning, 14*, 345-355.

[22]   Nielsen, K.: 1976, 'On Human Needs and Moral Appraisals', *Inquiry, 6*, 170-183

[23]   Rescher, J.: 1972, *Welfare: The Social Issues in Philosophical Perspective*, University of Pittsburgh Press, Pittsburgh.

[24]   Scanlon, T.M.: 1991, 'The moral basis of interpersonal comparisons', in Elster, J. & Roemer, J.E. (eds.), *Interpersonal Comparisons of Well-being*, Cambridge University Press, Cambridge, Massachusetts, pp. 217-244.

[25]   Sen, A.: 1985, 'Well-being, agency and freedom', The Dewey Lectures 1984, *Journal of Philosophy, 82*, 169-221.

[26]   Sundby, N.K.: 1974, *Om normer*, Universitetsforlaget, Oslo.

[27]   Tracy, L.: 1986, 'Toward an improved need theory, in response to legitimate criticism', *Behavioral Science, 31*, 205-218.

[28]   Tåhlin, M.: 1990, 'Politics, dynamics and individualism - the Swedish approach to level of living research', *Social Indiciators Research, 22*, 155-180.

[29]   Unesco: 1977, *Unesco's Policy Relevant Quality of Life Research Programme*, UNESCO, Division of Socio-economic Analysis, Paris.

[30]   Wahba, M.A,. and Bridwell, L.G.: 1976, 'Maslow reconsidered: A review of research on the need hierarchy theory', *Organizational Behavior and Human Performance, 15*, 212-240.

[31]   Wilcox, A.R.: 1981, 'Dissatisfaction with satisfaction: Subjective social indicators and the quality of life', in Johnston, D.F. (ed.), *Measurement of Subjective Phenomena*, Special Demographic Analyses, Dept. of Commerce, Washington, D.C.

[32]   Østerberg, D.: 1988, *Metasociology*, Norwegian University Press, Oslo.

DOES SELF-DECEPTION ENHANCE THE QUALITY OF LIFE?

In everyday life we present ourselves, our thoughts and feelings, to others. We may present our inner experiences as they are, as we perceive them, or as we wish others to perceive them. In principle, we are the only ones who know them. This gives us a free hand.

Self-presentations are sometimes referred to as "masks". Santayana (quoted by Goffman [15]) regards self-presentations as protection, necessary for survival:

Masks are arrested expressions and admirable echoes of feeling.... Words and images are like shells, no less integral parts of nature than are the substances they cover, but better addressed to the eye and more open to observation.

In quality of life research people present their feelings to the researcher. It is generally accepted that deception is widespread in self-reports on quality of life. In such contexts deception is considered a methodological problem, as discussed by Paulhus [32]. However, I shall not go into this here, as methodology is not the subject of this article.

In this article I shall describe and discuss the role of deception as a determinant of quality of life. If a woman declares herself to be a happy person — more happy than she actually is — what are the consequences? Will the statement make her more happy, or less happy? First of all, the consequences for herself as an individual will be discussed. But I shall also look into some possible effects on social relationships. And finally I shall consider some possible ideological consequences.

DECEPTION AND SELF-DECEPTION

'Deception' may be defined as "the act of representing as true what is known to be false", i.e., a deliberate attempt to mislead, manipulate or cheat another person. In everyday life we sometimes present ourselves to others with the intention to create a good image, a better image than we consider justified. This is a form of "impression management" [30] [31]. For example, we might tell our friends that we went on a ten-hour hike during the week-end, even though we did not walk for more than eight hours.

L. Nordenfelt (ed.), Concepts and Measurement of Quality of Life in Health Care, 95–114.

We also use the term 'deception' in compound words, such as 'self-deception'. de Sousa defines self-deception as follows:

...we may characterize self-deception ... as the purposive act of entertaining a belief that one has good ground for holding most likely false, in the service of some more comfortable state of mind [5].

According to Martin:

Self-deceivers ... refuse to acknowledge truths that they would recognize if they respected rational procedures [23].

When self-deception is involved, the intention is not only to mislead others, but also, and primarily, to mislead ourselves. It allows imagination to transcend reality. After having bragged about our ten-hour hike a couple of times, we begin to believe that the hike actually lasted for ten hours. We become convinced by our own lies. In experimental social psychology, Festinger induced subjects to tell lies as part of the experiment, and found that they came to believe these lies [11]. Moreover, we often elicit the cooperation of others in sustaining our self-deception. The social surroundings may support the self-deceiver out of politeness or because they are misled by the lie [36]. This sustention by social interaction is perhaps the rule, rather than the exception. Truth tellers, such as Gregers Werle in Ibsen's drama *The Wild Duck* or the child in H.C. Andersen's *The Emperor's New Clothes* are often considered socially awkward or naive.

Even though an act is intentional it is not necessarily conscious. A person who tells himself that he is more loving or more self-sacrificing than he actually is, may be deceiving himself. He may highlight his positive features and de-emphasize his flaws. He does so intentionally to enhance his self-esteem. But to say that he does so consciously would perhaps be a contradiction in terms. Or, perhaps not. Self-deception has been conceived as

... a conscious refusal to attend to unstructured tacit knowledge of which one is simultaneously aware [21].

In recent years, self-deception has been a recurring theme in philosophy and theoretical psychology [22] [23] [41] [46]. It has been argued that self-deception in any literal sense is impossible [23] [24]. According to Martin

...it is impossible to both know and not know the same thing at the same time in the same respect [23].

However, referring to Haight [18], Martin acknowledges that

...the term 'self-deception' can be metaphorically applied to forms of behavior that writers on self-deception have often had in mind. For example, 'self-deceivers' might know something that they are motivated not to become conscious of on occasion when recollection would normally be expected [23].

In the following, I shall use the term 'self-deception' in this loose, metaphorical sense. It is conceived as a species of deception addressed primarily to oneself. However, other-deception and self-deception are closely intertwined:

...self-deceivers often use deception, and deceivers are often self-deceived [36].

In this field it is, I believe, most fruitful to consider *degrees* of consciousness and *degrees* of intentionality.

The fact is we can be *somewhat* aware of intentions to conceal ourselves from ourselves ... [50].

Thus, a distortion may be more or less intentional and more or less conscious. It may be more or less tailored to an audience and more or less addressed to oneself.

Some degree of deception and self-deception is probably normal, e.g., to claim that we are not offended even though someone has attempted to insult us. Extreme degrees of honesty may be a rarity, perhaps even considered pathological. Slight exaggerations, hypocrisy, insincerity and small lies may be more healthy than scrupulous honesty.

Self-deception should be distinguished from the lowering of ambitions, which may be a conscious and realistic strategy in the face of a problematic situation. It also should be distinguished from Freudian repression, which is by definition unconscious [20]. Much in the mind is unconscious, but can easily be brought to consciousness, whereas repressed material cannot [29]. One may say that repression is a form of self-deception, but not all forms of self-deception are repression.

QUALITY OF LIFE

In this article the term 'quality of life' will be defined as positive minus negative affective experiences. Feeling sad, dejected, lonely are examples of negative affective experiences, whereas feeling happy, glad, enthusiastic, loving are examples of positive affective experiences. I do not use the terms 'positive' and 'negative' in an ethical sense, but to describe a feeling tone from the point of view of the experiencing person.

Most definitions of quality of life include not only affective experiences, but cognitive experiences as well. However, the above definition does not. Cognitive experiences, e.g., life satisfaction, will, as conceived by the author of this article, strongly affect the emotions. If we set ourselves particular short-term goals like catching the train, or long-term goals like winning the love of a certain person, the fulfillment of these goals will probably give us satisfaction and thus contribute to our positive affect. But if the goal fulfillment does not, for some reason or other, lead to positive affect, our quality of life is not enhanced. However, it will often be difficult to separate cognitive and affective experiences. I think it is safe to say that most cognitive experiences have emotional aspects and most affective experiences involve cognition.

At any rate, our topic is a psychological state, not a condition of life, such as a person's social or material circumstances. The terms 'psychological well-being' or 'emotional well-being' may be used alternately with the term 'quality of life' [1] [28].

DECEPTION AND QUALITY OF LIFE

In the case of quality of life self-reports, a person may give a false self-presentation, e.g., to the effect that he is more happy than he knows that he is. The fact that 'quality of life' as defined in this article relates to a person's state of mind makes it in principle inaccessible to others. The experiencing person is the final judge. In spite of this, we often feel that we are being given false self-presentations. Sometimes we feel that the person is not as happy as he says he is because he *looks* dejected. He *gives* the information that he is happy, but he *gives off* the information that he is not [16] [5]. Even if the person is smiling and looks cheerful, we may doubt the message. For some reason or other, we feel he is not really happy, he is just pretending. In other cases it is the other way around. We feel the person is *more* happy, or less sad, than he says he is. We think he is pretending to be sad because the situation calls for sadness. He may, e.g., have recently lost his father, for whom he did not care. He feigns sadness in order to fit the role of the bereaved son. A certain degree of hypocrisy, in this case to play sad, is appropriate to the situation [15].

Innumerable examples could be given of situations where deception is suspected. But in some situations the concept of deception is misleading. A person may express something about his feelings which is not true. But the statement is not interpreted as true, neither by the sender nor by the receiver. The conversation may simply be an exchange of pleasantries such as "How are you today? Fine, thank you. And you?"

In other settings, however, the interpretation may be more of a problem. When a group of persons is asked in a study about the quality of, e.g., their housing conditions, their marriages, their social status or popularity, their happiness or peace of mind, the general tendency is for the answers to gather in the positive end of the scale [4]. Thus most people seem to think that their happiness is greater than average. Now we have no a priori reason to presume that experiences such as happiness are normally distributed in our part of the world, or anywhere for that matter. It is possible, of course, that more people are happy than unhappy [43]. All the same, common sense tells us that reports of happiness are often exaggerated. We know that it is easy to confuse a realistic judgement with wishful thinking or intentional distortion of the actual state of affairs.

IDEALIZATION: POLLYANNA

This tendency to interpret one's own situation and report one's own experiences in an overly positive way may be called 'idealization' or 'idylization'. The phenomenon has also been called the 'Pollyanna syndrome'. Pollyanna was a heroine in a series of American children's novels published at the beginning of this century [34]. She preached the gospel of gladness, the virtues of positive thinking. The dear little orphan met with a series of misfortunes, but she had decided to look on the bright side of things, regardless of what happened to her. She invented the "just being glad game".

This game has been referred to as "the greatest game ever discovered since the foundation of the world", but also as a "tearjerker of a story", "sweet and sticky" [17].

The games we play are not restricted to the "being glad game". We also play the "being a martyr game" and the "being melancholy game". Shakespeare's much-quoted lines about the world being a stage, and "all the men and women merely players" has a more general application. But it seems that the "being glad" game is widespread. This article will not address the general phenomenon of playing games, but will discuss some possible causes and functions of the phenomenon of pretending to be glad.

The Pollyanna figure is a prototype so grossly exaggerated that she leaves us cold or disgusted. All the same, we are all to some extent natural Pollyannas. Everyone behaves like a Pollyanna under certain circumstances. That is, we sometimes pretend to be more glad, or less sad, than we actually are.

POSITIVE CONSEQUENCES

Why do we do this? What is to be gained by feigning and pretending? In the following I shall discuss some of the positive consequences of pretending to be more glad or less sad than we actually are.

According to socio-biological theories the capacity for deception and self-deception may offer a selective advantage:

...the ability to deceive increases fitness, and...self-deception increases the ability to deceive others [29].

...deceit and self-deception work hand-in-hand in helping individuals accomplish the critical task of becoming socially acceptable, thereby enhancing the likelihood of reproduction [9].

Thus, according to the socio-biologists, deception and self-deception support us in our attempts to attract the opposite sex. When we present ourselves as available and attractive partners, our reproductive success is enhanced by our ability to highlight our positive traits and disguise our negative ones. Cheerfulness may be considered a positive trait and quarrelsomeness a negative one.

In everyday life we sometimes observe that people refrain from describing their own situation in negative terms because they do not want to complain. They talk about pleasant subjects and keep unpleasant thoughts to themselves. They are afraid people will dislike them if they complain. This has received empirical support in experimental social psychology. College students were presented with pictures of unknown persons with varying emotional expressions. Asked to pick out the faces they liked, the students showed strong preferences for faces showing positive emotions and viewed the negative emotions with disapproval [40].

The tendency to refrain from complaining applies particularly to those in a weak position, such as sick and old persons. They have a strong need for positive regard and experience social pressure to bear their pain stoically. Thus they suppress unsuitable affect [15]. Consider a lonely woman of 75, living on a small pension in a run-down flat in a desolate part of town. She cannot afford to move to a better place to live. Nor can she afford to irritate the people around her by lamenting and groaning. She knows complaining will not make her popular, as most people like to be surrounded by happy faces. In the words of the poet Wilcox:

Laugh, and the world laughs with you.
Weep, and you weep alone;
For the sad old earth must borrow its mirth,
But has trouble enough of its own [49].

The strength of this norm probably varies from culture to culture and it also varies with the circumstances. According to anthropologist Wikan, this norm is stronger among the Balinese than in our culture [48]. The Balinese believe that the feelings we express influence the feelings of others. "When we see another sad, we may become like that too" Like moral acts, emotions can be chosen. Therefore emotion work is a collective duty.

Popular health care in Bali is thus first and foremost a matter of composing one's emotion by a deliberate effort of thought and expression and contributing to the maintenance of a cheerful and smooth social ambience [48].

Reports from other parts of the world reveal similar attitudes. Donner [7] relates an incident observed in an Indian tribe in Venezuela. A young man broke his leg and gasped in pain as a cast was being put on it. His sister and young wife "giggled" each time he moaned. "They were not amused but were trying to cheer him up."

The Western attitude towards suppression of feelings is ambiguous. In psychotherapy it is considered "healthy" to uncover layers of suffering, to express negative feelings, to "flood out". But as indicated by the above poem, positive expressions are more welcome in everyday life. Furthermore, positive self-presentations may enhance the sender's status. To be happy is a sign of success, not necessarily worldly success, but success as related to some personal goal. We tend to admire persons who appear satisfied with life. However, sometimes we do not admire, but rather envy them, which means that one should be careful not to over-state one's happiness.

One may say that happy self-presentations are socially desirable. They are addressed to an audience, to others, and the purpose is to be seen in a more favorable light. The lonely woman presents herself as a friendly, brave and contented person and is respected.

In competitive sport, the players may employ a deceptive strategy. Whittaker-Bleuler observed players in tennis matches and concluded:

Deception, particularly in the form of masking losing behavior, may be an important strategy for the championship athlete playing against a skilled opponent. By masking losing affect, submissive behavior is suppressed and the sender displays an aura of confidence [47].

The two norms *Don't complain!* and *Be successful!* and their relation to expressed quality of life are presented in Figure I. They are examples of deliberate attempts to control the information we pass on to other people.

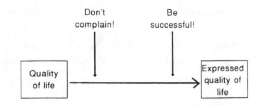

FIGURE I. Norms affecting the expression of quality of life

What we gain from this kind of behavior seems to be that we please other people and cut a good figure. We may also please and make a good impression on ourselves. This is where self-deception comes in. Our presentation of our own quality of life may enhance it; it may improve our self-esteem. We tell ourselves that we are happy, and this makes us happy or, rather, it makes us somewhat more happy or less unhappy than we were before. Self-deception can be used for purposes of gaining pleasure as well as avoiding pain [33].

Empirical evidence shows that self-deceit characterizes the non-depressed person rather than the depressed. A negative correlation between scores on self-deception questionnaires and depression measurements has been cited as evidence that "it is the non-depressed individuals who exercise more distortion" [35]. Similar results have been obtained in relation to psychopathology scores [37], insecurity and authoritarianism [24] [27] [31], self-dissatisfaction [38] and anxiety [38] [44]. These results may be given various interpretations. They may support the view that the psychopathology inventories are invalid [37]. However, the rival view that self-deceptive persons are more healthy than those who are less self-deceptive has also been launched [27].

Feedback effects

In "facial feedback studies" it has been found that when subjects are instructed to act out particular emotions, this tends to affect subsequent self-reports of their emotional states [51]. This feedback effect is illustrated in Figure II. We have put in a second arrow, signifying that the report is to be considered not only an expression, but also a cause.

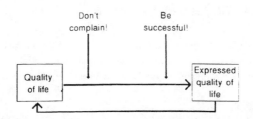

FIGURE II. The feed-back effect on expressed quality of life

A happiness report may be oral, or it may be transmitted by body language. When we smile and laugh and have a cheerful expression on our face, this will be interpreted as a sign, an indicator, of a good mood. But laughter is not only a sign, it is also a cause of a good mood. According to Wikan, the Balinese hold that:

Laughter ... creates health and happiness. ... laughter works even on unwilling minds... And when one wills oneself to laugh, the gain is manifold as the body is swept up in a surge of good energy flow [48].

The idea behind this is that emotions can be chosen at will. There are rules and conventions, not only of appearances, but also of feelings themselves. de Sousa calls this "ideologies of emotions" [5]. Thus it is assumed that emotions are responsive to deliberate attempts to suppress or evoke them.

Emotion work

This contention supports the interactive view of emotions. It contradicts the organismic notion that feelings are mainly instinctive and impulsive. According to the organismic view, the *expression* of emotion is subject to social influence, but the emotion *itself* is regarded as unmanageable, like a shiver or blushing.

Hochschild regards emotional management, or emotion work, from an interactive perspective. She proposes a distinction between emotion work and control or suppression:

The latter two terms suggest an effort merely to stifle or prevent feeling. 'Emotion work' refers more broadly to the act of evoking or shaping, as well as suppressing, feeling in oneself [20].

She also distinguishes between various techniques of emotion work:

One is *cognitive*: the attempt to change images, ideas, or thoughts.... A second is *bodily*: The attempt to change somatic or other physical symptoms...(e.g., trying to breathe slowly...). Third, there is expressive emotion work: Trying to change *expressive* gestures...(e.g., trying to smile...).

The cognitive approach is readily recognizable. It seems that it is most frequently used as a means to keep painful emotions at a distance, to make life tolerable. In self-deception we demonstrate diminished recall of negative evaluative characteristics [26]. Our old woman is worried firstly because the apartment house she lives in may be condemned, and her landlord, in antic-ipation of this, is negligent as to the upkeep of the building. Secondly she is worried because she has not heard from her daughter for a fortnight. She tells herself that she is not seriously worried or badly hurt, just a little. She practices selective attention and selective inattention. She compares herself with other persons who are clearly less fortunate than she is. She perceives, recognizes, remembers and communicates pleasant events, and ignores or forgets unpleasant ones. She lowers her ambitions to make the gap between expectation level and reality as small as possible [25]. As she cannot have another apartment, she decides that it is not worth the pursuit. Her ambition once was to go and live with her daughter. But she does not admit that her hopes cannot be realized; instead she claims that she does not want to. Elster has called this "sour grapes" or, more generally, "adaptive preference formation" [8].

But is this deception? Not necessarily. Lowering of ambitions may be conscious and realistic. However, claiming that the original goal is not worth the pursuit (sour grapes) may be deceptive. Further, attempts to impress other people by deliberately putting our best foot forward are of course deceptive. Our not so good foot is kept in obscurity. And most of the time when we work on our emotions, there is an element of self-deception involved. Sometimes, but not often, we start out with a very rational, conscious plan for improving our well-being. However, to bring the action to a successful end may be difficult without a good portion of self-deception. It does not work if we do not believe in it.

Self-deception has been studied as a coping strategy in threatening situations. In the face of a death threat, such as a cancer diagnosis, some patients have been found to deny or minimize the threat. In a recent study threat *minimization* was compared with other coping strategies, such as searching for affiliation, searching for information, searching for meaning in religion. Threat minimization was found to be one of the most successful strategies in altering levels of well-being [12].

COULD OR SHOULD DECEPTION BE ENCOURAGED?

We have discussed some positive consequences of posing as happy: The poser may avoid negative reactions from other persons by keeping his sorrows to himself. He may win the respect of others by appearing successful. And last but not least, he may change his own feelings in a positive way, that is actually *feel* more happy, e.g., by telling and convincing himself and others that he is in fact rather happy, by playing the glad game.

But does this mean that we should encourage deception? In what way *could* we encourage deception? Like the Balinese, we could socialize our children to control their bad and sad moods. We could train them in emotion work [48]. In psychotherapy we could teach patients various techniques of emotional management. Admittedly, most people do not see a psychologist or a psychiatrist. But the general practitioner receives patients with emotional problems every day and has to choose between various therapeutic approaches.

Fordyce proposes a "program to increase personal happiness". He reports a series of studies in which college students participated in courses teaching them fourteen "fundamentals". Several of these may be considered to be cognitive techniques, such as "stop worrying", "lower your expectations", "develop positive, optimistic thinking", "eliminate negative feelings and problems", "put happiness as your most important priority". Fordyce concludes:

there exists a growing, research-based knowledge of happiness...this knowledge can be used to help individuals better realize their pursuit of happiness [14].

Cognitive therapy is based on the notion that the content of a person's thoughts is crucial to his emotional experiences.

It simply means that we get to the person's emotions through his cognitions [3].

Rumination is considered not only a symptom, but also a cause, a reinforcer, of depressive thoughts. "Adaptive self-talk" or "self-instructional training" is incorporated into therapeutic techniques [19].

It seems safe to conclude that deception can be encouraged, directly or indirectly. Certain conditions, certain cultures, are supportive of distortion, others counteract it. However, this is not the same as saying that deception *should* be encouraged. In the following, I shall discuss some arguments against what may be called "the culture or ideology of deception".

NEGATIVE CONSEQUENCES

We may distinguish between negative consequences for the posing individual himself and negative consequences involving a wider circle. In the following I shall discuss a number of arguments that have been put forward against the notion that deception has positive effects.

The social relations argument

It may be argued that deception is a barrier to human communication, to empathy with others. Deception may be desirable in brief encounters "by contributing to a cheerful and smooth social ambience" (as Wikan puts it in her comment on the Balinese way of interacting [48]). However, deception disrupts *close* interpersonal relationships.

Close relationships constitute the most important factor in the quality of most people's lives [1] [4] [2]. Deception is destructive, as openness is crucial in such relationships; it is the basis of trust and love.

The old woman's daughter may be confused when she visits her mother. She may also feel somewhat ashamed because she has not visited her for a long time. But her mother does not complain in spite of the daughter's negligence, and in spite of the bad state and uncertainty relating to her apartment. Is she really content, or is she concealing her true feelings? Maybe she does not care, as she puts on the same smile, regardless of what happens. Or is she trying to save face?

This may be the reason why the "positive thinking" praised in the figure of Pollyanna is not accepted without reservation in our culture. In some settings she is considered foolish and contemptible. The very obviousness of the role playing subjects her to social ostracism. However, as we have pointed out, the complainer is not a popular figure either. The ideal of most people is perhaps better expressed in the following comment on the Pollyanna figure:

The correct response is not to be a Pollyanna who thinks everything is 'peachy-keen' nor a 'Melancholy Baby' but someone in between — 'a levelheaded realist' [17].

In some situations involving personal crises, a realistic attitude is of particular importance, e.g., when problem drinkers or gamblers are urged to stop. It has been argued by Fingarette that the medical community's designation of alcoholism as a disease tends to encourage the individual to deny responsibility. Thus the self-deception inherent in alcoholism is actually promoted [13]. It serves as an excuse, intended to absolve the agent of any blame for his condition [39]. If the person in question is willing to face the gravity of the situation, it is more likely that he will benefit from support and help from close friends and relatives.

We tend to feel contempt for the role-playing person. Self-deception is considered debasing, degrading and maladaptive because it distorts reality. Self-insight, i.e. the opposite of self-deception, is proclaimed as a Socratic virtue. The norm *Know yourself!* is widely accepted.

Elster [8] presents a similar ideal in his denouncement of the "sour gapes" reaction. He proposes an alternative which he calls "character planning". This approach presupposes insight and deliberate choice. "It is better to

adapt to the inevitable through choice than by non-conscious resignation."
Instead of downgrading inaccessible options ("sour grapes"), deliberate
character planning would tend to upgrade accessible ones. Conscious
lowering of ambitions is another approach. The ideal is to pursue the good
life in a realistic manner.

To sum up the social relations argument, the enhancement of quality of
life through deception may yield short-term benefits. However, the long-
term benefits are questionable, as deception seems to be detrimental to close
social relations.

The value argument

Truth is given higher priority than quality of life in the value systems of
some philosophers, e.g., Kant, Kierkegaard, Nietzsche and Heidegger [23].
Self-deception is considered morally onerous. According to Kant, self-deception is "a contemptible violation of human dignity" [23].

In the words of one Norwegian philosopher:

... we are confronted with the chasm between an authentic life worthy of man...and an illusory
life lived in pleasant self-deception.... The choice implies the unconditional acceptance of the
value of human dignity at the cost of traditional objectives such as adjustment, success,
happiness, peace of mind, etc. [42].

These philosophers hold that the question whether deception enhances
quality of life is irrelevant. Authenticity, or human dignity, is the highest
value, irrespective of the consequences for psychological well-being. Enhancing the personal insight of human beings is the only acceptable purpose of
scientific and philosophical endeavor.

The emotional costs argument

When people seek happiness, some desire to be happy most of the time, even if only mildly
so, whereas others appear to live and plan for rare but intense moments of ecstasy [6].

The conditions that reduce unhappiness may be different from those that
enhance high levels of happiness. The strategies involving deception discussed above may be effective in relation to episodes of serious unhappiness.
Moreover, they may be effective in raising moderate levels of happiness to

somewhat higher levels. But they may also be ineffective, or even contra-indicated, when the goal is to achieve intense moments of highly positive affect.

Why is this so? It has been argued by Diener *et al.* [6] that intense positive experiences are quite rare, and that they may frequently have undesirable side-effects. These aspects tend to offset the benefit of intense positive emotions. Some studies indicate that factors which intuitively lead to intense positive affect can also heighten the intensity of negative affect.

A number of interrelated factors which influence the intensity of positive emotions also influence the intensity of negative emotions for the individual: high assessment of the importance of events; repression; cognitive amplifying and dampening strategies; and physiological reactivity. The end result of these mechanisms is the same: to amplify or dampen both positive and negative affective responses. Thus it seems that in the long run in peoples lives, many high peaks will be paid for to some extent by lower lows when the person becomes unhappy [6].

This may mean that the mechanisms involving deception which aim at redu-cing pain may inhibit or serve as a barrier to peak experiences. The emotion work, the cognitive strategies we use, may alleviate suffering but at the same time be a hindrance to intense positive experiences. Such experiences involve emotional costs. Cognition which allows high peaks presupposes or leads to more extreme lows.

If this is true, we may have to choose. One option would be an emotional life consisting of both intense negative and intense positive experiences. Another option would be an emotional life consisting of a fairly smooth level of feelings.

According to the study by Diener *et al.* [6], a person's tendency to express happiness is determined by the frequency, not the intensity, of his emotional experiences. Diener's conclusion is that quality of life researchers should operationalize happiness in terms of frequency, rather than in terms of intensity.

This seems to speak in favour of accepting some measure of self-decep-tion. Such a conclusion is, of course, not warranted. The fact that this is the choice of some (or most) American research subjects does not mean that it should be set up as an ideal for everyone and for other cultures.

Some cultures and some individuals give priority to intense, though rare, moments of happiness; others choose more frequent, though milder, positive experiences. Value judgements play an important role in this. Empirical facts are not in themselves decisive.

The social change argument

The above discussion focuses on the enhancement of personal happiness. But the enhancement of individual happiness will sometimes be in conflict with the common good. It may be an obstacle to social improvement.

A person playing the "glad game" under adverse circumstances will not be inclined to try to change these circumstances. Our old woman does not want to complain. She is afraid that her daughter will not respond well to criticism. She tries to adjust, to suppress her indignation and stress the positive aspects of her situation. She claims that her landlord has the right to do as he pleases with his own property. As a person he is benevolent and doing his best. As for her daughter she is sometimes very helpful and supportive in spite of being very busy. These thoughts cheer her up. They take the edge off her distress.

But this also does away with her motivation to change the situation. She is not likely to tell her daughter that she needs more attention. The other tenants are organizing a protest, but she will definitely not join forces with them to stop the landlord's plans.

According to critical theory, she has been socialized by means of a dominant ideology to accept a submissive role in her relationships to other people. To use the Marxist terminology, she does not recognize her real interests; she is the victim of false consciousness [10].

Both her daughter and the landlord may more or less consciously employ a strategy that keeps her reasonably content while at the same time taking care of their own interests.

This woman definitely derives short-term benefits from deceiving others and herself as to her well-being. Her long-term well-being may be more questionable. Furthermore, she certainly does not contribute to the common cause of improving the well-being of her fellow tenants in their run-down apartment house.

It is not difficult to find more dramatic examples of self-deception. In our part of the world we enjoy a standard of living which, if shared by the rest of humanity, would have catastrophic consequences. The injustices as well as the predicted global changes resulting from our Western way of life do not seem to impress us in any perceptible way.

CONCLUSION?

Sorry, I have no conclusion. Or, rather, my conclusion is a trivial one: Sometimes, in some situations, both self-deception and other-deception enhance quality of life, in other situations they do not.

A more interesting question is whether deception should be encouraged. I am reluctant to provide an answer. The value argument presented above is not one I support. The emotional costs argument is also, in part, a value argument, and is more in accord with my value system. The other arguments are of an empirical nature. Is it true that deception is destructive to close relationships? Or, rather, in which situations is it destructive? Is it true that false consciousness is advantageous in the short run, but not in the long run? To whom? Under what circumstances? These questions are challenging, and the need for more research is clear.

BIBLIOGRAPHY

[1]   Andrews, F. M. and Robinson, J. P.: 1991, 'Measures of Subjective Well-Being', in J. P. Robinson, P.R. Shaver and L.S. Wrightsman (eds.), *Measures of Personality and Social Psychological Attitudes*, Volume 1, Academic Press, San Diego.

[2]   Argyle, M. and Martin, M.: 1991, 'The psychological causes of happiness', in F. Strack, M. Argyle and N. Schwarz (eds.), *Subjective Well-Being*, Pergamon Press, Oxford.

[3]   Beck, A.T.: 1976, *Cognitive Therapy and the Emotional Disorders*, International Universities Press, New York.

[4]   Campbell, A., Converse, P. E. and Rogers, W.L.: 1976, *The Quality of American Life*, Russell Sage Foundation, New York.

[5]   de Sousa, R. B.: 1988, 'Emotion and self-deception' in McLaughlin, B.P. and Rorty, A.O. (eds.) *Perspectives on Self-Deception*, University of California Press, Berkeley.

[6]   Diener, E., Sandvik, E. and Pavot, W.: 1991, 'Happiness is the frequency, not the intensity, of positive versus negative affect', in F. Strack, M. Argyle and N. Schwarz (eds.), *Subjective Well-Being*, Pergamon Press, Oxford.

[7]   Donner, F.: 1984, *Shabono*, Triad/Paladin Books, London.

[8]   Elster, J.: 1987, *Sour Grapes*, Cambridge University Press, Cambridge.

[9]   Essock, S.M., McGuire, M.T. and Hooper, B.: 1988, 'Self-deception in socialsupport networks' in Lockard, J.S. and Paulhus, D.L. (eds.) *Self-Deception: An Adaptive Mechanism?*, Prentice Hall, Englewood Cliffs, New Jersey.

[10]   Eyerman, R.: 1981, *False Consciousness and Idiology in Marxist Theory*, Almqvist & Wiksell, Stockholm.

[11]   Festinger, F. and Carlsmith, J.M.: 1959, 'Cognitive consequences of forced
         compliance', *The Journal of Abnormal and Social Psychology*, 58, 203-210.

[12]   Filipp, S.-H. and Klauer, T.: 1991, 'Subjective well-being in the face of critical life
         events: the case of successful copers', in Strack, F., Argyle, M. and Schwarz, N.
         (eds.) *Subjective Well-Being*, Pergamon Press, Oxford.

[13]   Fingarette, H.: 1985, 'Alcoholism and self-deception', in Martin, M.W. (ed.) *Self-
         Deception and Self-Understanding*, University Press of Kansas, Lawrence.

[14]   Fordyce, M.W.: 1983, 'A Program to Increase Happiness: Further Studies',
         Journal of Counseling Psychology, 30, 483-498.

[15]   Goffman, E.: 1961, *Encounters*, Bobbs-Merrill, Indianapolis.

[16]   Goffman, E.: 1980, *The Presentation of Self in Everyday Life*, Penguin Books,
         Harmonsworth.

[17]   Griswold, J.: 1987, 'Pollyanna — ex-bubblehead', *New York Times Book Review*,
         (Oct.), p. 51.

[18]   Haight, M.R.: 1985, 'Tales from a black box' in Martin, M.W. (ed.), *Self-
         Deception and Self-Understanding*, University Press of Kansas, Lawrence.

[19]   Hawton, K., Salkovskis, P.M., Kirk, J. and Clark, D.M. (eds.): 1989, *Cognitive
         Behaviour Therapy for Psychiatric Problems*, Oxford University Press, Oxford.

[20]   Hochschild, A.R.: 1979, 'Emotion work, feeling rules, and social structure',
         American Journal of Sociology, 85, 551-575.

[21]   Joseph, R.: 1980, 'Awareness, the origin of thought, and the role of conscious self-
         deception in resistance and repression', *Psychological Reports*, 46, 767-781.

[22]   Lockard, J.S. and Paulhus, D.L. (eds.): 1988, *Self-Deception: An Adaptive
         Mechanism?*, Prentice Hall, Englewood Cliffs.

[23]   Martin, M.W.: 1985, 'General introduction', in Martin, M.W. (ed.), *Self-Deception
         and Self-Understanding*, University Press of Kansas, Lawrence.

[24]   McCrae, R.R. and Costa Jr., P.T.: 1983, 'Social desirability scales: More substance
         than style', *Journal of Consulting and Clinical Psychology*, 51, 882-888.

[25]   Michalos, A.C.: 1985, 'Multiple Discrepancies Theory (MDT)', *Social Indicators
         Research*, 16, 347-413.

[26]   Millham, J. and Kellogg, R.W.: 1980, 'Need for social approval: Impression
         management or self-deception?', *Journal of Research in Personality*, 14, 445-
         457.

[27]   Monts, J.K., Zurcher, L.A. and Nydegger, R.V.: 1977, 'Interpersonal self-
         deception and personality correlates', *The Journal of Social Psychology*, 103, 91-
         99.

[28]   Naess, S.: 1987, *Quality of Life Research*, Institute of Applied Social Research,
         Oslo.

[29]   Nesse, R.M. and Lloyd, A.T.: 1992, 'The evolution of psychodynamic mechanisms', in Barkow, J., Cosmides, L. and Tooby, J.(eds.), *The Adapted Mind*, Oxford University Press, New York.

[30]   Paulhus, D.L.: 1984, 'Two-component models of socially desirable responding', *Journal of Personality and Social Psychology* 46, pp. 598-609.

[31]   Paulhus, D.L.: 1986, 'Self-deception and impression management in test responses' in Angleitner, A. and Wiggins (eds.), *Personality assessment via questionnaire*, Springer-Verlag, New York.

[32]   Paulhus, D.L.: 1991, 'Measurement and control of response bias', in J.P. Robinson, P.R. Shaver and L.S. Wrightsman (eds.), *Measures of Personality and Social Psychological Attitudes*, Volume 1, Academic Press, San Diego.

[33]   Paulhus, D.L. & Reid, D.B.: 1991, 'Enhancement and denial in socially desirable responding', *Journal of Personality and Social Psychology*, 60, 307-317.

[34]   Porter, E. H.: 1975, *Pollyanna*, Penguin Books, Harmondsworth.

[35]   Roth, D.L. & Ingram, R.E.: 1985, 'Factors in the Self-Deception Questionnaire: Associations with depression', *Journal of Personality and Social Behavior*, 48, 243-251.

[36]   Ruddick, W.: 1988, 'Social self-deceptions' in McLaughlin, B. and Rorty, A.O. (eds.), *Perspectives on Self-Deception*, University of California Press, Berkeley.

[37]   Sackeim, H.A. and Gur, R.C.: 1979, 'Self-deception, other-deception, and self-reported psychopathology', *Journal of Consulting and Clinical Psychology*, 47, 213-215.

[38]   Schwartz, D.B.: 1982, 'Effects of short-term self-awareness training on self-deception, reflective-ability, anxiety and self-dissatisfaction', *Dissertation Abstracts International*, Ann Arbor, 43:1 80A, Jul.

[39]   Snyder, C.R.: 1985, 'Collaborative companions: The relationship of self- deception and excuse making', in Martin, M.W. (ed.), *Self-Deception and Self-Understanding*, University Press of Kansas, Lawrence.

[40]   Sommers, S.: 1984, 'Reported emotions and conventions of emotionality among college students', *Journal of Personality and Social Psychology*, 46, 207-215.

[41]   Steffen, L.H.: 1986, *Self-Deception and the Common Life*, American University Studies, Peter Lang, New York.

[42]   Tennessen, H.: 1966-1967, 'Happiness is for the pigs: Philosophy versus psycho-therapy', *Journal of Existentialism*, 7, 181-214.

[43]   Veenhoven, R.: 1991, 'Questions on happiness: classical topics, modern answers, blind spots', in F. Strack, M. Argyle and N. Schwarz (eds.), *Subjective Well-Being*, Pergamon Press, Oxford.

[44]   Vitaliano, P.W., Katon, W., Russo, J., Maiuro, R.D, Anderson, K. & Jones, M.:
         1987, 'Coping as an index of illness behavior in panic disorder', *The Journal of
         Nervous and Mental Disease*, *175*, 78-84.

[45]   Welles, J.F.:1988, 'Societal roles in self-deception', in Lockard, J. S. and Paulhus,
         D.L. (eds.), *Self-Deception: An Adaptive Mechanism?*, Prentice Hall, Englewood
         Cliffs.

[46]   Whisner, W.: 1989, 'Self-deception, human emotion, and moral responsibility: To-
         ward a pluralistic conceptual scheme', *Journal for the Theory of Social Behaviour*,
         19, 389-410.

[47]   Whittaker-Bleuler, S.A.: 1988, 'Deception and self-deception. A dominance
         strategy in competitive sport', in Lockard, J.S. and Paulhus, D.L. (eds.), *Self-
         Deception: An Adaptive Mechanism?*, Prentice Hall, New Jersey.

[48]   Wikan, U.: 1989, 'Managing the Heart to Brighten Face and Soul. Emotions in
         Balinese Morality and Health Care', *American Ethnologist*, *16*, 294-312.

[49]   Wilcox, E.W.: 1980, 'Solitude', in J. Cooper (ed.), *Violets and Vinegar*, London.

[50]   Wilshire, B.: 1988, 'Mimetic engulfment and self-deception' in McLaughlin, B. and
         Rorty, A.O. (eds.), *Perspectives on Self-Deception*, University of California Press,
         Berkeley.

[51]   Winton, W.M.: 1986, 'The role of facial response in self-reports of emotion: A
         critique of Laird', *Journal of Personality and Social Psychology*, *50*, 808-812.

SECTION II

QUALITY OF LIFE IN THE HEALTH
CARE CONTEXT: ANALYTICAL AND
ETHICAL ISSUES

MICHAEL BURY

QUALITY OF LIFE: WHY NOW?

A SOCIOLOGICAL VIEW

INTRODUCTION

The last few years have witnessed the renewal of considerable debate about the character, definition and measurement of quality of life. This has been especially true with respect to the application of concepts of quality of life in health assessments and health care evaluation. However, in Britain, at least, accounts of the background and reasons for this increase in interest in quality of life are less in evidence. Much of the current debate focuses on difficulties in defining and deciding what are the relevant dimensions of quality of life for particular purposes, and of operationalizing these in research [15]. There is also considerable debate on the use, or more properly, possible use, of quality of life measures in the policy sphere, particularly in the allocation of resources, again, most especially in health care. This latter debate has included ethical considerations, and the possible unintended consequences of the practical application of quality of life measures. For example, there are fears that people could be excluded from receiving care if their membership of a disease group, rated low in terms of quality of life, was used as a means of rationing [22].

The purpose of this chapter is to take a step back from these debates, and consider a broader question, namely why quality of life issues have now come to the fore in health care, particularly in research and in policy circles. My treatment of the subject is necessarily tentative at this stage, given the shifting sands of contemporary concerns. I am concerned here with sketching the contexts within which recent interest in quality of life measures have developed, rather than attempting to assess possible causal influences on their emergence, though I may seem to imply certain "effects" of these contexts at various points.

Three broad areas will be addressed. First, there is the most immediate and obvious context of the changing health and demographic profiles of "late modern" societies. In this context, I want to identify some of the limits of traditional approaches to health and longevity that have emerged in the last twenty years. This first part of the discussion, focuses, therefore, on key features of the changing nature of contemporary experience and its impact

117

L. Nordenfelt (ed.), Concepts and Measurement of Quality of Life in Health Care, 117–134.
© 1994 Kluwer Academic Publishers. Printed in the Netherlands.

on our thinking. Second, there is the "political economy" context of quality of life measures, especially in relation to health care policy. This part of the discussion will consider, in broad terms, the "restructuring of welfare" currently under way in a number of contemporary societies, and the ideological attraction of quality of life measures in such a context. Third, and last, I want to consider the wider cultural context in which quality of life measures have arisen. At this point it will be necessary to tackle the argument that quality of life measures, and research, are essentially part of an expanding field of expertise that extends the jurisdiction of professional power and surveillance in everyday life. A consideration of the strengths and weaknesses of this argument will conclude the chapter.

CHANGING PATTERNS OF HEALTH

I wish to begin, however, with a consideration of changing patterns of health and health disorders, and the social contexts in which they are occurring. A way into this set of issues, following Rosenberg [32], is what can be thought of as the "frames and framers" of disease. Rosenberg's basic argument is that, while the "frames" placed round health and disease are human and, therefore, social products, these, in turn, are affected by the nature of disease experience and its impact. For example, poor child health and maternal mortality, in the first part of this century, were the subject of much debate and controversy, but the content and characteristics of the experiences involved themselves influenced the concepts and frames employed. Nutrition, for example, was the focus of much argument because of its potentially crucial role, at that time, in the health of mothers and babies. Though such factors could not determine the outcome of the debate (was deprivation or "poor parenting" to blame?) they could place limits on what was argued about.

The importance of seeing health experiences, and the frames which are developed to encapsulate them, as interacting, is underlined, by Rosenberg, by considering the effects that specific features of disease or disorder can have on human responses. For example, evidence about the route of transmission of infectious disease has had a major impact on social responses, as the history of diseases such as cholera illustrate [12] and as AIDS has done in more recent times. It is plain that responses to AIDS have been shaped by the fact that one of the key routes of transmission is held to be through sexual contact. Though "frames" around such disorders are not simple reflections of an a-historical reality, nor based on infallible knowledge, neither can their development and change be seen simply as an arbitrary social product. I shall return to the implications of this point in the last section of this chapter.

For the moment, I simply wish to take up Rosenberg's argument in the context of changing health patterns and their influence on current thinking about quality of life. Though a concern with such matters as life threatening disorders among children and young people remains socially significant, any assessment of health disorders in contemporary Western society is confronted with the secular decline in mortality rates, as well as in long-term trend towards lower fertility rates, and the associated, and widely recognized ageing of the population that has accompanied them. An ageing population brings with it, in turn, a relative increase in the dominance of chronic disorders which influence both "frames and framers" in health care. Whilst the nature of the responses to such changes are not wholly determined by them, it remains the case that they are framed in a context less dominated by the infections and more and more by the diseases and disorders of later life. Moreover, the expansion in the range of effective treatments mediates these developments [19] extending the lives of people of all ages who would earlier be subject to fatal disease. As a result, the impact of medicine is increasingly less likely to be seen in terms of survival alone.

Traditionally, public health has been concerned, not to say preoccupied, with mortality. Public health "frames" in the first half of this century were developed and articulated to help cope with the complex patterns of "premature" mortality, and, to a lesser extent, the incidence and prevalence of morbidity. Fitzpatrick and Dunnel [15] point up the dominance of premature and avoidable death as measures of health in most public health and clinical research in the twentieth century. Such preoccupations, however, lose their grip on the consciousness of modern populations and are, to some extent, displaced towards the problems of survival in developing countries, most especially Africa. The relative security of life in "late modern" societies means that reality of death appears to be remote, especially to the young, hidden behind the veil of "civilized" life [20]. Elias argues that this makes our inevitable confrontation with dying and death particularly problematic, though he is quick to note that the decline in mortality has had the positive effect of releasing people from the ever present fear of untimely death [11]. Perhaps it is also important to add that age specific mortality rates, especially among unskilled working class males, do not always confirm this otherwise secular trend [30]. Moreover, deaths from AIDS, again, has reminded us of the power of infectious disease to shape our frames of reference. The effects of this disease have recently been shown to be affecting mortality rates among young men [10].

Nevertheless, key changes in social experience and health disorders remain important in their impact on our thinking. Illsley, for example, has summarized the significance of changes this century by highlighting reductions in the proportion of all deaths occurring in early life, especially from the "classical poverty diseases of infection and respiratory disease." Deaths under the age 15 fell from 27% of all deaths in 1921 to only 4% in 1971. At the same time, only 32% of males died at age 65 or over in 1921, whereas by 1971 this figure had risen to 65% [23]. Today, over ninety per cent of all mortality in Western societies occurs over the age of 55. It is something of a paradox, as Patrick points out, that as mortality rates improve and as recording and understanding progresses, they become less relevant as measures of health or as measures of the outcome of medical intervention [27].

As I have indicated, the earlier concerns with infectious disease and mortality are not only superseded by the problems associated with improved survival, but also by problems associated with the unintended consequences of medical treatments. The growth of effective treatments transforms many conditions that may have been fatal into chronic ones, particularly as age barriers to treatment fall, as is the case in Britain. The examples of the expansion of treatment for renal failure, heart disease and some forms of cancer are obvious ones. In many chronic disorders people have to adapt not only to illness, but also to managing long-term treatment regimens [34]. Living with the disabling effects of such disorders and their treatment, brings the ability to function and live in daily settings into focus. It is clear from this that the orthodox model of disease, in which curative treatment was the key goal, loses its hold over thinking and practice.

However, it is important to note, here, that these historic changes in population age structures and associated morbidity patterns should not be seen as a linear process of an inevitable growing burden of chronic illness. As Fries has argued, if the age of onset of chronic disease could be postponed by preventive health measures and judicious use of health sevices, morbidity could be "compressed" into the last years if not months of life, rather than adding years of morbidity to the existing level. Though evidence for this occurring at present remains contentious [33] [17] [5] quality of life measures could, with improving health, be a means of documenting positive change, and the capacity of people to adapt, as much as of extending the documentation of a growing burden of chronic illness associated with an ageing population.

In either case, the arguments of Patrick [28] and others, in favor of an extension of the orthodox model to encompass the consequences of chronic disorders, in a "socio-medical" model, to take into account measures of disability, handicap, well-being and life satisfaction, may be seen as "frames" placed round current experiences. Though the terms inevitably involve "second order" constructs, their value is bound to be tested against the "first order" experiences which they purport to address. The earlier framework of the "natural history" of disease in terms of aetiology, pathology and manifestation of symptoms simply does not offer the necessary "coverage" for researchers, health care providers, or indeed, people in general, where "management" and "adaptation," rather than cure, now prevails. The consequences of disease and health disorders over time bring wider considerations into view.

A brief comment on a recent study of the very old, involving the present author, may help to illustrate the point [7]. In this national study of the health and social circumstances of people over ninety years of age in England, we found it difficult to avoid the issue of quality of life, both in thinking about our respondents' experience across the life course and in our research design, even though we were somewhat unsure of the meaning of the term to the very old. As I shall examine below, this situation may be a testament to the power of fashions in modern thought, of modern "discourses" in our thinking, as much as the effects of changing realities, but the inadequacy of simply documenting morbidity levels, or even disablement, was evident from the outset (see also Fallowfield [13] on the relevance of quality of life measure for the elderly). It was clear, for example that morbidity levels could not always predict people's subjective responses or preoccupations, including, critically those concerning social relationships.

We developed our assessments of quality of life from the dimensions suggested by George and Bearon [18] for studies in old age. Four dimensions are involved: health and functional status; socio-economic status; life satisfaction; and self esteem. Such dimensions can be applied to both individuals, and to others, such as family members or carers. In the case of very old age, the family context is crucial given the distance such people are from their working lives and work settings. Family care, dependency, and questions such as morale and loneliness emerged, as well as autonomy and risk taking, as of central importance. The study emphasized that morbidity and disablement measures alone also fail to capture the outcomes of interactions between current circumstances and experiences across a long life course.

Space permits just one example. The study found considerable variation among responses to questions concerning aspects of current life satisfaction. When asked whether the people in the study looked forward to the day on waking, the results showed that the majority did. Table one summarize the findings.

TABLE I. Looking forward to the day on waking

	%
Usually	65.1
Sometimes	6.6
Indifferent	17.6
Rarely	5.5
Never	5.2

Though there were significant associations with factors such as degree of mobility restriction and pain, these accounted for only a minority in the sample. The majority looked forward to daily life, often despite the presence of restrictions and poor health. Factors such as incontinence, often thought to mean poor quality of life, failed to predict these sorts of measures. Moreover, the presence of good relationships with family acted as a buffer between circumstance and low self-esteem ([7], pp. 111-112). In this study, a quality of life perspective helped to capture such interactions, especially if they were set in a framework of the life course.

To summarize, the purpose of the discussion in this section has been to suggest that the concern with quality of life concepts and their operationalizing in research, stems from real changes in experiences, linked to an ageing population and to the growing importance of chronic illness, for individuals, carers and providers. This "realist" note may offset a tendency among some commentators to account for changes in modes of thought *only* in terms of rather arbitrary "cultural constructions." My argument is that the terms and concepts (or "frames") involved in the "quality-of-life" debate need first to be seen as emerging in response to changing realities, and the *social experiences* of health to which they relate, in late twentieth century life.

THE POLITICAL ECONOMY OF QUALITY OF LIFE

Second, and alongside these developments, is the changing economic and political context, especially as it has affected the provision of health care. From the time of Cochrane's and McKeown's critiques, in the nineteen seventies [8] [25] the impact of an expanding range of treatments to more diverse groups of patients has been seen to be in need of more systematic evaluation, in terms of efficiency and effectiveness. Soon the issue of "value for money" became linked to this challenge. Health care had become both more extensive and expensive. Interestingly, the development of concern with evaluating the role of curative medicine came originally, not from the neo-conservative Right but from the reforming Left of social policy, in alliance with public health specialists. From this viewpoint it was difficult to separate new ways of assessing health experiences from assessments of the institutional forms developed ostensibly to promote them.

From the outset, therefore, challenges to a reliance on the relatively insensitive indicators of mortality rates have been made by a variety of interests outside the clinical arena. As economists, policy analysts and politicians all voiced concerns about the future of health and welfare systems, alternatives, such as quality of life measures, have become caught up in assessing the performance of services as well as with the assessment of individuals (see for example [1]).

The interest in effectiveness and efficiency of health services has stemmed in part, in Britain at least, from a long standing set of conflicts between community (now public health) medicine and clinical medicine. Challenges to orthodox medical definitions of health, and the need to study the effectiveness of treatment modalities, were, in Cochrane or McKeown's arguments, put forward against the putative dominance of an unevaluated acute sector in health care provision. The concern with efficiency, on the other hand, was expressed with an eye on the public purse, and a possible dialogue with politicians, who were, from the nineteen seventies onwards, increasingly keen to place limits on public spending. By challenging the medical establishment in this way community medicine, in particular, was attempting to find a new direction away from its traditional role in local authority public health surveillance of infectious disease [24]. McKeown, for example, argued that orthodox "mechanical" medicine was persisting with notions of "cure," when good quality *care* was needed, especially for the elderly and chronically sick. In this way it was hoped that community medicine, reorientating itself, could be seen as being at the forefront in tackling the need for change in the health service.

The search for a new identity in community medicine opened up possibilities of collaboration with a growing medical sociology (for example, Sir Douglas Black's collaboration with Peter Townsend on the "Black Report," [37]), and with health economists. Throughout the period, clinical medicine was somewhat slow to realize the growth of debates about effectiveness, partly as a result of its preoccupations with technical advance and its traditional reliance on clinical judgement and individualized treatment. Early programs of transplant and by-pass surgery, for example, relied on evidence of reduced mortality, expressed in terms of individuals achieving two or five year post-operative survival. When survival appeared poor, clinical medicine was put on the defensive. Cochrane's well known critique of the treatment of coronary heart disease, and its lack of comparison with other forms of treatment and use of control groups, is perhaps the best known example in Britain [8]. Moratoria on transplant programs in the U.K. and the U.S. were dramatic results of criticism of the lack of scientific evidence of clinical effectiveness [16].

As I have indicated, the tension between social and clinical medicine soon became caught up in the wider transformations of health and welfare systems at the end of the 1970s and beginning of the 1980s. In fact, Cochranes's book appeared at the turning point of this period — both economically and politically — 1972. The year was characterized by fiscal crisis and the first major post-war attempt to limit public expenditure, by a Labour government under pressure from the IMF [14] [21]. The evaluation of health care, therefore, financially, as well as in terms of the outcomes of treatment, steadily became a central political issue, and a major part of subsequent conservative policy in Britain. The 'new right' rather than the left were now in the policy driving seat. This has been taking place in many European countries, including Sweden, which, until recently, has seen its welfare system as immune from such developments. Such political change gave added weight (and from a quarter not foreseen by earlier reformers) to arguments that had hitherto been concerned with the desire to challenge the supposed lack of proven effectiveness of modern medicine. Cochrane's hope that randomized control trials (RCTs) could sort out good from bad practice, in terms of rational science, was soon to be overshadowed and overtaken by considerations of a much wider nature.

In Britain, at least, the restructuring of welfare, and particularly health care that was involved, seems to have gone through two broad stages during this period. According to Pollitt [29] the first concerned itself with value for money and increasing efficiency, as attempts were made to drive down, or hold down, public spending. During the first half of the 1980s, this largely involved a confrontational stance by politicians against professional groups, including doctors. Ironically, the work of community medicine and medical sociology, in identifying the lack of an unequivocal relationship between the provision of health care and an improvement in health status, had laid some of the grounds for the suggestion that more expenditure would not necessarily mean better health (at least, when measured in terms of average life expectancy).

British conservative governments turned away from professionals towards businessmen and business techniques, as a way of tying in the welfare state more tightly to the economy, and the efforts to reduce inflation. As health service administration was being turned into a more sharply focused management [9], rewards and sanctions attached to strict budgeting overshadowed other considerations, including achieving more appropriate outcomes of care. In as much as "quality" (of care) was used it meant such items as developing data on "performance indicators," such as length of stay in hospital, waiting lists and other measures of "throughput," coupled with the beginnings of a recognition of consumer views.

In fact, Pollitt [29] argues that despite a program of budgetary limits and cuts in NHS services, as elsewhere in the public sector, a reduction in overall levels of public expenditure was not produced (though the *rate* may have been slowed). What was produced, however, was a considerable lowering of morale. The attempts to use (in the NHS) new management structures and techniques, in which performance was to be gauged by achieving lower spending targets, and greater productivity, ran into numerous difficulties, including professional interests (particularly surrounding the exercise of clinical autonomy) and the limits of the politics of confrontation.

Pollitt argues that the second stage, which took off in 1986/7, was marked, in contrast, by an emphasis on "human resource management" and a set of quasi-scientific decision-making techniques to achieve "quality," though the term seemed to have almost limitless meaning. However, attempts were made to consider the outcomes of care based on wider definitions of health status (other than merely recovery or survival), and quality of life. Discussion of the Quality Adjusted Life Year (QALY) has perhaps been the most extensive. The debate about QALYs, and their possible use, can therefore be seen as part of the second wave of managerialism that developed from the mid 1980s onwards.

Against the language of (clinical) needs or "needology" as it is sometimes disparagingly referred to, a new language emerged. Quality Adjusted Life Years, comprising life expectancy following treatment, weighted to take account of disability levels and subjective distress, became an attractive talking point for the new managers, especially when they were combined with the costs of specific procedures. Not surprisingly, therefore, health economists developing these measures became increasingly influential, as compared with community physicians and medical sociologists, in their earlier attempts to broaden health status measures. Taking the argument that resources are finite, and "needs" unlimited, a language appealing to notions of effectiveness in terms of quality of life, particularly when it could be combined with a commitment to cost containment in the NHS, rapidly developed as an "empowering" managerial ideology.

Quality of life measures, in this context, soon moved from being research issues for economists and others, to being explored by managers and the new specialists in public health medicine (the successor to community medicine) as potential guides to health policy. Instead of providing more resources to meet needs, a better quality of service could be aimed for, within properly managed budgets. Instead of clinicians rationing by restricting treatment to individuals, or groups such as the elderly, new priorities might now be set (at least theoretically) with QALY-type measures helping to define "best buys." Work on surgical programs such as hip replacement and coronary by-pass seemed to point in this direction [38]. Quality of care could now be couched not only in terms of *process* (features of services such as waiting times) but also *outcome*, in terms of quality of life. If "quality" could also be translated into consumer satisfaction, the full spectrum of service delivery could carry a "quality" dimension. After all, as with sin, everyone is opposed to services of poor quality.

This helps to explain, perhaps, the wide use of the term 'quality' by politicians and economists in referring to innumerable features of the welfare restructuring process. Pollitt [29] points out, however, that this process is not simply an ideological tool in a new managerialism in the NHS though it clearly is serving this purpose. The use of performance and quality measures, including quality of life measures, appears to provide managers with an important "window" onto providers' activities, and acts as a check on professional autonomy. What needs to be noted, however, is that the rapid interest in such measures has the hallmark of a crisis reaction. Research efforts to explore the possibilities and limitations of such measures, especially in the assessments of health amongst older people and those with chronic illnesses [13] has been overshadowed by work on the outcomes of acute services, especially surgery, with the much more controversial aim of influencing the allocation of resources. It is this process which has led to the allegation that quality of life measures, such as QALYs, have taken on a reified character, turning complex issues of experience and meaning into numerical expressions, which then have a life of their own. This is even of more concern when it is realized that the early research by Rosser and others that paved the way for QALYs, was based on very limited numbers of respondents, many of whom were, in fact, professionals [3].

However, though there is much talk of such measures, as a means of reconciling expenditure and outcome problems, there is little evidence of their regular or operational use ([29], p. 4). The apparently compelling nature of quality of life measures, from this viewpoint, suggests the need for combing a political economy approach with a further, and somewhat more complex, level of analysis.

QUALITY OF LIFE AND "LATE MODERNITY"

Having located the development of interest in quality of life within the contexts of changing health experience and the restructuring of welfare, there is, then, a third issue that needs to be addressed. This is that the massive expansion of information and research of the type discussed here (and quality of life measures have to be numbered alongside a variety of other socio-medical and psycho-social indicators, see Bowling [4]) can be seen, from a sociological viewpoint, as part and parcel of the rapid transformations in the institutions and everyday life settings of late modern society. Here, key features of late modern cultures, alongside health experiences or political economy, need to be brought into focus, taking us back to issues introduced earlier in this chapter. A central part of the argument, at this level, is with the consequences of the move away from concerns with survival to a concern with "standards of living." With the increasing domination of society over nature, threats from the natural world, including those of disease, become transposed into fields of social action. As we have seen, preventing disease and premature death, and effecting cures, are superseded by concerns with social circumstance and the adaptation to the effects of chronic conditions, especially in later life [6]. Social relationships and their maintenance become as important as the treatment of symptoms.

In fact, information about, and surveys of, the quality of life, have accompanied the development of modern society from its earliest years. Surveys of the conditions of urban dwellers abound in all modern societies, certainly from the nineteenth century onwards, and have often been important, and contentious, markers in the political life of modern nation states. Health status, housing and other amenities, not to mention disposable income and wealth, have all figured in such studies.

Najman and Levine [26] discuss the rapid development of quality of life measures in the post World War II context. As with today, these have been marked by debate and dispute. They isolate four main problems which have dogged the heels of ideas about quality of life from their inception.

First, Najman and Levine argue that conceptualizing what constitutes quality of life has often been vague. This has meant that agreement about what adds up to a good or poor, high or low, quality of life has been minimal. Second, disagreement about the use of specific indicators has often meant that different surveys (especially over time) could not be compared. Third, there have been problems in relating "inputs" to "outputs" in terms of quality of life. Initiatives in social or welfare policy (including, perhaps, developments in health services) might well have been associated with changes in quality of life indicators, but it has been difficult to know whether such "outcomes" have been the result of the initiative or service, or of some other variable (e.g. economic change). Current research on quality of life faces similar problems; showing that a specific change in quality of life is causally linked to a specific service requires tackling formidable methodological problems, especially those of the role of "spurious variables." Showing that survival alone has improved, or has been achieved, as a result of treatment, is relatively straightforward by comparison. Fourth, there have been particular difficulties relating objective measures to subjective values or assessments. For example, George and Bearon, in the study of quality of life and old age (mentioned earlier in this chapter [18]), make the point that income clearly has a bearing on quality of life, but that older people may value more income less than, say, maintaining health.

Far from resolving these methodological and value issues, it might be argued that the "discourse" on quality of life has subsequently grown like topsy, moving into areas of subjectivity and social relationships, where measurement becomes ever more problematic, and where valuations may prove to be highly context dependent. Indeed, quality of life measures (at least those used in the socio-medical field) may become so preoccupied with such matters as "well-being" and "self esteem" that little reference to "objective" or material contexts occurs at all. This may be seen to be especially prevalent among some psychologists (for a discussion see [13]). In seeking increasingly to "understand" an ever wider range of the subjective aspects of human life. Professional discourse, from this viewpoint, effectively abstracts the person from real-life circumstances.

This argument would suggest that quality of life measures are, therefore, part of a wider attempt to provide "cultural constructions" of "the whole person" [2]. Following Foucault, Armstrong, for example, argues that the tendency of modern society increasingly to survey and construct subjectivity under the guise of welfare, acts as a powerful "disciplinary" code or "regime of truth" (see also [31]). The huge expansion in quality of life measures (along with numerous other techniques and sources of information about people's lifestyles and beliefs, from political polls to consumer preferences) is held to mark something more than the needs of either a changing health scene, or a rising managerialism. A deeper process is alleged to be at work, in which people's subjective views are constantly caught up in a web of surveillance.

Past and present concerns with quality of life might be understood, from this viewpoint, as being central to the operation of power in modern social systems; providing information about people's subjective views them but also playing a key role in constituting their very character. As Armstrong puts it: "In the past it was sufficient to monitor and control the body... now the very core of self must be rendered docile" ([2], p. 32). Passivity may result, precisely because expertise cloaks itself in a language reflecting consumer rather than professional values. After all, how can decisions about prioritizing resources be opposed if they are based on the valuations of life gained from surveys of the subjective views of ordinary people?

There is much in this argument from a sociological viewpoint. A critical sociology must necessarily draw attention to the effects, intended and unintended, of developments in professional expertise and practice (including both medicine and the social sciences) which are advanced under the name of extending "concern" and welfare. Not only this, but it is part of the critical role of sociology to attempt to enquire behind the facade of "contemporary history" to provide an interpretation of "why now?" which many advocates of new modes of thinking themselves fail to provide. Quality of life measures may be advocated as better ways of comprehending experience or being more "rational" in the use of resources, but they also may be serving vested interests, in health care and in the wider political context. Moreover, the language of "rationality" itself may serve to disguise its own contradictions, masking the fact that today's "rational" measure (or measurement) is tomorrow's forgotten fashion. At the very least, scientific modes of thought need to be recognized as producing only provisional truths, capable of being revised with new knowledge. The idea that modern thought is a "long narrative" of accumulating certainties, and that measures such as quality of life ones provide a certain foundation for decisionmaking, needs to be approached with caution.

Having said this, however, there are, in my view, serious weaknesses in the argument that concepts such as 'quality of life,' and the measurements attempting to operationalize them, are best understood as little more than components of surveillance-producing "discourses." There are three brief points I would like to make, by way of a conclusion, in this connection.

First, critiques of the power of experts, tend to treat each development in thinking as if it is equally effective in producing a powerful language or "discourse" over everyday modes of thought. Every development in knowledge or practice among professional circles (especially the human sciences and medicine) is simply then included in the same argument. Each development is treated as if it contains the same amount of surveillance potential, or the same degree of ideological power. This, at worst, tends to suggest that people outside of expert systems are "cultural dopes," unaware of the interests new knowledge may be serving. As Giddens has recently argued, however, the "sequestration" of everyday meanings by expert systems is not as complete and as negative as is sometimes alleged [20].

Quality of life measures, for example, may well reproduce lay experiences and subjective judgements in quasi-technical language, but this, in turn, may be used by lay actors to further their own interests. "Re-skilling" as well as "de-skilling" occur [20]. The essentially "contestable" character of knowledge production and assimilation in late modern societies warns against accepting the assumption that the colonizing of everyday experiences is either always successful, or complete. Thus issues such as quality of life become areas of argument and debate, both within professional circles and between professionals and lay groups, just as earlier versions of standards of living, mentioned above, were. Under conditions of "late modernity" expertise and information about human experience move back and forth between segments of the cultural order. Predictions about the outcome of such processes, especially that they will always produce more "docility" amongst the public are simply not warranted. New challenges to the traditional power of medicine may result. For example, calls by managers for the use of quality of life measures to allocate resources in the NHS have been contested by clinicians and patient groups, amongst others. The fact that such measures may be largely rhetorical, making little impact on actual decision-making or policy, suggests at the least that the power of "discourse" should not in any event be taken at face value.

Second, though the critics of quality of life measures suggest that these are part of the intellectual armor of the powerful, the argument actually turns on a weakly specified notion of power. In particular, the allegation that professional interests call forth these new types of "subjective" measures, often fails to recognize intra-professional disputes which fracture powerful groups. Intra-professional conflicts are as characteristic as lay/professional clashes. Moreover, the supposed "constitutive" properties of such expertise are only vaguely sketched out in such arguments. Thus, it often seems that vested interests and no specific interests are served by turns, as it is difficult to locate exactly who benefits from surveillance activities. This weak specification of power constitutes one of the most serious problems with the Foucauldian approach to modern cultures [35]. At a more mundane level, it is difficult to see how specific groups of lay actors, such as the elderly or those with chronic sickness, suffer from the ill effects of too much surveillance through the employment of quality of life measures. It is just as arguable that such lay groups have historically suffered from *neglect* as much as from the negative consequences of professional power or medicalization. The use, and possible misuse, of quality of life measures need not lead to their dismissal outright, but to an engaged argument about their meaning and employment.

Finally, explanations for the rise of quality of life measures, which rely on the supposed negative tendencies of modern (or "postmodern") social systems, often fail to specify why specific concepts appear to have an "elective affinity" with current experience. It can be argued that modern society's preoccupations with subjectivity and personal wellbeing may constitute a unifying as well as fragmenting force. Taylor [36] for example, locates modern concerns with "ordinary life," and specifically with the problems of production and reproduction (work and home life) within the context of the waning of religious values and their superseding by respect for the individual and human rights. In this sense, the preoccupations with quality of life and wellbeing can be seen as part of a secular change in modern social structures, in more ways than one. These developments are not simply ideological edifices, but pathways through the hazardous changes of late twentieth century life, though they are contestable at each stage. Taylor argues that frameworks governing everyday life are indispensable to modern attempts to address moral concerns. Here, the final answer to "why now?" is not simply, as Foucauldian thought suggests, merely a question about the growth of disciplinary power, but also the indispensable need (however faltering and at times contradictory) to address matters of human respect and autonomy.

Insofar as a concern with quality of life expresses the essential reflexive character of modern social systems, as they attempt to address questions of how we should live in everyday life, reactions need not be wholly negative [20]. The argument developed in this chapter suggests that whilst a critical view of the emergence and use of such measures is warranted, we should not see this as a one-way process of the loss of power and autonomy of the individual. The importance of measures such as those of quality of life, and indeed the reasons for their rapid development now, also reflect major changes in social experience, linked, as I have shown to changes in demography and the changing pattern health disorders. An agenda for sociology which offers nothing more than an overgeneralized critique runs the risk of leaving itself on the margins. A more engaged position, in which empirical research and development of the measures runs alongside the continuing argument about their employment and use, by lay groups (including patients) as well as policy makers and managers, would seem to suggest a more satisfactory way forward.

BIBLIOGRAPHY

[1]   Abel-Smith, B.: 1976, *Value For Money in the NHS*, Heinemann, London.

[2]   Armstrong, D.: 1986, 'The Problem of the Whole Person in Holistic Medicine', *Holistic Medicine, 1*, 27-36.

[3]   Ashmore, M., Mulkay, M. and Pinch, T.: 1989, *Health and Efficiency: A sociology of health economics*, Open University Press, Milton Keynes, U.K.

[4]   Bowling, A.: 1991, *Measuring Health: A review of quality of life measuring scales*, Open University Press, Milton Keynes, U.K.

[5]   Bury, M.: 1992, 'The Future of Ageing: Changing Perceptions and Realities', in J.C. Brocklehurst *et al.* (eds.), *Textbook of Geriatric Medicine and Gerontology*, Churchill Livingstone, London.

[6]   Bury, M.R.: 1991, 'The Sociology of Chronic Illness; A Review of Research and Prospects', *Sociology of Health and Illness, 13*, 451-468.

[7]   Bury, M.R. and Holme, A.: 1991, *Life After Ninety*, Routledge, London.

[8]   Cochrane, A.: 1972, *Effectiveness and Efficiency: Random Reflections on Health Services*, Nuffield Provincial Hospitals Trust, London.

[9]   Cox, D.: 1991, 'Health Service Management - a sociological view: Griffiths and the non-negotiated order of the hospital', in J. Gabe, M. Calnan & M. Bury (eds.), *The Sociology of the Health Service*, Routledge, London.

[10]   Dunnel, K.: 1991, 'Deaths among 15-44 year olds', *Population Trends*, *64*, 38-43.

[11]   Elias, N.: 1985, *The Loneliness of the Dying*, Basil Blackwell, Oxford.

[12]   Evans, R.: 1989, *Death in Hamburg*, Oxford University Press, Oxford.

[13]   Fallowfield, L.: 1990, *The Quality of Life: The Missing Measurement in Health Care*, Souvenir Press, London.

[14]   Fitzpatrick, R.: 1987, 'Political Science and Health Policy', in G. Scambler (ed.), *Sociological Theory and Medical Sociology*, Tavistock, London.

[15]   Fitzpatrick, R. & Dunnel, K.: 1992, 'Measuring Outcomes in Health Care', in Beek, *et al.*, *The Best of Health*, Chapman & Hall, London.

[16]   Fox, R.: 1974, *The Courage to Fail: A Social View of Organ Transplants and Dialysis*, University of Chicago Press, Chicago.

[17]   Fries, J.: 1989, 'The Compression of Morbidity: Near or Far?', *The Milbank Quarterly*, *67*, 208.

[18]   George, L.K. and Bearon, L.B.: 1980, *Quality of Life in Older Persons: Meaning and Measurement*, Human Science Press, New York.

[19]   Gerhardt, U.: 1989, 'Introductory Essay: qualitative research in chronic illness: The issue and the story', *Social Science and Medicine*, *30*, 1149-59.

[20]   Giddens, A.: 1991, *Modernity and Self-Identity: Self and Society in the Late Modern Age*, Polity Press, London.

[21]   Le Grand, J.: 1982, *The Strategy of Equality: Redistribution and the Social Services*, Unwin Hyman, London.

[22]   Harris, J.: 1988, 'Eqalyty' in P. Byrne (ed.), *Health, Rights and Resources*, King Edward's Hospital Fund, London.

[23]   Illsley, R.: 1987, 'The Health Divide: Bad welfare or bad statistics?', *Poverty*, *67*, 16-17.

[24]   Jefferys, M.: 1986, 'The Transition from Public Health to Community Medicine: The Evolution and Execution of a Policy for Occupational Transformation', *The Society for the History of Medicine, Bulletin*, *39*, 47-63.

[25]   McKeown, T.: 1976, *The Role of Medicine: Dream Mirage or Nemesis*, Nuffield Provincial Hospital Trust, London.

[26]   Najman, J.M. and Levine, S.: 1981, 'Evaluating the Impact of Medical Care and Technologies on the Quality of Life: A Review and Critique', *Social Science and Medicine*, *15*, 102-16.

[27]   Patrick, D.: 1986, 'Evaluating Health Care', in D. Patrick and G. Scambler (eds.),
        Sociology as Applied to Medicine, Bailliere Tindall, London.

[28]   Patrick, D.L. and Peach, H. (eds.): 1989, *Disablement in the Community*, Oxford
        University Press, Oxford.

[29]   Pollitt, C.: 1991, 'The Politics of Quality: Managers, Professionals and Consumers in
        the Public Services', lecture for the Centre for Political Studies, Royal Holloway and
        Bedford New College, University of London.

[30]   Powles, J.: 1978, 'The Effects of Health Services on Adult Male Mortality in Relation
        to Social and Economic Factors', *Ethics, Science and Medicine*, 5, 1-13.

[31]   Rose, N.: 1990, *Governing the Soul: the shaping of the private self*, Routledge,
        London.

[32]   Rosenberg, C.E.: 1989, 'Disease in History: Frames and framers', *Milbank
        Quarterly*, 67, Suppl. 1, 1-15.

[33]   Schneider, E.L. and Brody, J.A.: 1983, 'Aging, Natural Death and the Compression
        of Morbidity; another view', *New England Journal of Medicine*, 309, 854.

[34]   Strauss, A. & Glaser, B.: 1975, (2nd ed, 1984), *Chronic Illness and the Quality of
        Life*, Mosby, St Louis.

[35]   Taylor, C.: 1986, 'Foucault on Freedom and Truth', in D.C. Hoy (ed.), *Foucault: A
        Critical Reader*, Basil Blackwell, Oxford.

[36]   Taylor, C.: 1989, *Sources of the Self: The Making of the Modern Identity*, Cambridge
        University Press, Cambridge.

[37]   Townsend, P. and Davidson, N. (eds.): 1982, *Inequalities in Health: The Black
        Report*, Penguin Books, Harmondsworth, U.K.

[38]   Williams, A.: 1985, 'Economics of Coronary Bypass Grafting', *British Medical
        Journal*, 291, 326-329.

ANNE FAGOT-LARGEAULT

REFLECTIONS ON THE NOTION OF 'QUALITY OF LIFE'[1]

The notion of "quality of life" is very much in the air these days. The reflections that follow consider it from the medical point of view ("health-related quality of life": HQL). Let me offer an example. Concerning the treatment of chronic diseases it is no longer enough to prove that a new therapy is effective and non-toxic; it has to be proved that in the case of efficacy and toxicity comparable to those of standard treatment, the new treatment gives the patient an improved quality of life. Thus, for instance, arterial hypertension (AHT) can easily be controlled by a variety of drugs that have little toxicity at effective dosage levels but have side-effects (nightmares, diminished sexual capacity, depression) which are often so difficult to bear as to cause the abandonment of treatment. During the past ten years, great effort has been devoted — with the aid of the pharmaceutical industry — to attempt to ascertain what therapies offer a technically satisfactory treatment of AHT without involving a deterioration in quality of life [see, e.g., 11].

I begin by observing that there is nothing new about introducing considerations of quantity/quality of life to aid in resolving a medical "dilemma". What is relatively new, in democratic societies, is the effort to make these considerations explicit, to argue them publicly and to justify the choices that have to be made in terms of the number and value of the years of life "gained" by means of a particular therapeutic strategy. Not everyone appreciates this type of calculation. Putting it broadly, the aversion to the sort of argument involving quantity/quality of life seems to denote a preference for a deontological ethics (morality based on duty), whilst recourse to such argument denotes a preference for a utilitarian ethics (morality with a teleological foundation, oriented towards the maximum good or happiness). I shall first follow the utilitarians in defining good action, or the ethical decision, as that action/decision which maximizes a quantity adjusted by a quality. I shall show that utilitarian theorists, having valiantly set off down the road of calculative rationality, have made manifest the paradoxes and aporias to be encountered at the end of this road, not only in the domain of health but elsewhere. This encounter has caused their confidence in rationality to be tempered by scepticism. I turn to the deontologists, who assume the principle of the "sacred" character of human life, which completely excludes any calculation whereby one life is taken to have more or less value than another. However, we shall see that, be it from compassion or from a sense of realism, the supporters of "absolute" respect for life surreptitiously reintroduce in practice the case-dependent nuances they reject in theory. Yet even though in fact there finally turns out to be an overlapping of the two types of solution offered, it remains true that there are two distinct attitudes with regard to the confronting of difficult choices. This it is important to bear in mind. The fact is, procedural solutions that involve the possibility of arriving at decisions by

135

L. Nordenfelt (ed.), Concepts and Measurement of Quality of Life in Health Care, 135–160.
© 1994 Kluwer Academic Publishers. Printed in the Netherlands.

way of negotiation without taking a position regarding fundamentals, remain a source of unease especially if the parties confronting each other do not grasp their lack of agreement.

Jay Katz tells the story ([31], p. 5) of a "distinguished French nephrologist" who (at the beginning of the 60s, probably), having diagnosed chronic renal failure in a patient who resided in the country, explained to him the inexorably progressive nature of the disease, and then let him return home. Katz, who had a conversation with the nephrologist soon after the patient left, was deeply upset by this, and he asked the doctor why he had not mentioned the possibility of iterative chronic haemodialysis. The doctor replied that it would have been cruel to mention it. To receive such treatment the patient would have to move near a large hospital — but "peasants do not adjust well to a permanent move to a large city". It was the doctor's opinion that for this man a shorter survival in his familiar surroundings was preferable to a longer survival in alien surroundings. Katz does not dispute the rightness of giving precedence to the *quality* of survival rather than to its duration, he simply regrets that the patient was given no opportunity to express his preference. (Today in France, both the patient and his family would have been included in the decisionmaking process.)

Medical practice has always involved dilemmas, tragic or painful choices one cannot escape. In the resolution of such dilemmas doctors and midwives have always tacitly introduced, either in the recesses of their own consciences or in their conversations with the families, a notion of value or *quality of life*. Take for instance the case of a difficult birth. Whose life is of greater value, the mother's or the child's? Or, take the case of a new-born child with multiple anomalies. Is resuscitation worthwhile? In the days when the majority of deliveries occurred at home employing manual techniques, whereby the doctor or midwife had the chance to give nature a helping hand rather than substituting for her, the decisions were made on the spot, and those who made them disliked speaking about it — the rules of professional practice were transmitted by example, with many options remaining tacit. Now that births take place under collective care, in hospitals that have at their disposal a rich array of technical resources — where it is possible to keep premature babies alive who in the past would not have survived, to artificially provide for the natural deficiencies of the new-born child, and to surgically correct grave deformities — such dilemmas as whether to proceed with or to abandon the resuscitation of severely underweight newborns come into prominence for nursing teams, families and, because of the high cost of resuscitation, for hospital administrations [34] [58]. The situations have be-

come codified, the prognostic criteria have been clearly defined, and deontological rules have emerged [12] [21]. This evolution is clearly irreversible. Though it is possible to be nostalgic about the way things were,
when the majority of people were unaware of the difficulties and it appeared
that choices virtually made themselves, there is no going back to such an
"ignorance condition". To set "practices policies" [14] involves an attempt
at analyzing and rationalizing medical practices, and even when such policies finally turn out to confirm intuitions about what should be done, there
is no skimping on the work of rationalization.

The process of rationalization is manifest in a second series of examples
that concern the care lavished not at the beginning of life but towards its
close [35]. Because a growing proportion of persons die in hospitals, in hos-
pices ("long stay" service) or in homes for the elderly, the management of
dying is less and less a private matter [52] [22]. The institutions where
people die must adopt a care policy since they are accountable to families as
well as health insurance companies. Nowadays there are powerful technological means for delaying death: peritoneal dialysis, cardiorespiratory resuscitation, and organ transplantation to name a few. These interventions are
introduced even in cases of patients who are very old and/or very ill (e.g.,
suffering from metastatic cancer or Alzheimer's disease). Can such persons
be denied the benefit of these medical interventions [50]? Persons nearing
the end of life are often lucid enough to express a preference, but that does
not resolve all the problems. How should the institution respond to a request
that treatment be discontinued (or continued indefinitely), or, indeed, a request for euthanasia? Or, if the person is not lucid, or the questions have not
been posed, what attitude should the institution adopt if a serious medical
problem should arise? Take for instance the issue of whether a systematic effort should be made to resuscitate everyone who suffers a first cardiac arrest.
Must there be entered in the dossier of certain pensioners, as is done in
North America, an order not to resuscitate? One of the ends of medicine has
always been the saving of lives. But there comes a borderline (rendered
more conspicuous by technological possibilities) beyond which this goal
ceases to have any meaning — either because it is judged "disproportionate"
to invest sophisticated medical resources in the prolongation of lives too
poor in quality, or because the scarcity of technological resources makes it
necessary to choose who is to benefit: two cardiac arrests at the same time,
one defibrillator available....

Michael Lockwood has pointed out that such a situation of scarcity can
be illustrated through a dilemma dramatized by George Bernard Shaw at the

beginning of the 20th century [57]. A research team are experimenting with a treatment designed to cure tuberculosis. They can treat n individuals, they have recruited n-1, there is one place left. Two candidates appear, a morally depraved genius and a decent but untalented man. Which of them should be chosen? The better one. But which one is that?

Since there is a right to health, correlative with a duty of the State to ensure that all persons have access to care, and to ensure that health resources are equitably distributed, the fact that these resources are limited (even though they absorb an important part of the GNP: 10% in the USA, 7-8% in Western Europe, 5.6% in the UK) makes it necessary to consider questions of justice on a global basis, and makes it necessary to attempt to optimize the use of the resources available and avoid squandering scarce resources. In the UK, where the economies imposed on and by the National Health Service (NHS) have forced a sharpening of awareness, renal dialysis has been made dependent on an age limit in the majority of public health institutions, although this limit has not been openly defended by argument. Giving dialysis at public expense to a person over 65 was generally considered a waste of resources. In France the increase in awareness proceeds more slowly. The socialist government in the late 1980s tried stirring public reflection through large forums at which people were supposed to say what sort of health coverage they wanted, and at what cost to them [Etats généraux de la santé]. Those forums aroused little interest, for at the same time politicians (and doctors) claimed that no one could ever be denied the medical care he/she needed. Still vividly remembered is the general outrage caused by the publication in the press of a "secret" circular of the Ministry of Health limiting the proportion of dialysed patients to receive an expensive new drug (erythropoietin). Neither the secret nor the outrage constituted evidence of political maturity — for it is clear that no health system has unlimited resources.

To recapitulate, there have always been medical dilemmas — to some extent bound up with the limits of medical knowledge and capacity, but not completely. For a long time these dilemmas have been resolved discreetly and intuitively, but today — because of increased technological possibilities, the collective nature of the decisions, and the need to control rising health care expenditures — they tend to be thrust into the public domain, and intuitive solutions no longer appear satisfactory.

The beginnings of modern medical rationality coincided (in the 18th and 19th centuries) with an effort to reason concerning medical decisions in terms of *quantity of life gained* ([18], Ch. 6). From the individual point of view the quantity of life is its duration, measured in number of years, whilst from the collective point of view the quantity of life is the average length of life of a generation, or life expectancy. When Daniel Bernoulli (1760) discussed the advantages and drawbacks of inoculation with variolar pus in order to prevent smallpox, he calculated that if everyone were inoculated the average length of life would increase from 26 years and 7 months to 29 years and 9 months, involving a gain of more than three years. He concluded that inoculation is the right choice, on both the individual and the collective level — in spite of the risk (not forgetting the by no means negligible risk inherent in inoculation as such). Some years later, thanks to the work of Jenner, the technique has improved: vaccination (inoculation with discharge from cowpox) has taken the place of inoculation with discharge from smallpox. There is less risk. Duvillard (1806) calculated that if everyone were vaccinated there would be an increase in average life expectancy from less than 29 years to more than 32. He concluded that vaccination should be carried out.

Thus it is possible to measure the advantage of a medical strategy by its calculated consequences, and with the admission that "more is better" the best collective strategy is the one that leads to the best increase in the mean duration of life. It was on the basis of this type of argument that doctors and hygienists in the 19th century first made an assault on infectious diseases — first and foremost those affecting children, the conquest of which provides great gains in life expectancy. There was here a rational order of priorities. After almost every possible gain had been made on this front (global European mortality went down by half in the last twenty years of the 19th century, whilst that of persons over 50 remained stable), in the 20th century an assault was made on afflictions of persons of mature age (work or road casualties, alcoholism, broncho-pulmonary cancer), the containment of which provides a less striking but nevertheless appreciable increase in life expectancy (which between 1780 and 1980 was tripled in Europe). At present it is hoped to gain additional years by making an assault on the diseases of old age, but the *problem of quality* then becomes crucial. In the developed countries it is no longer enough to gain years of life, there has to be a gain in quality of life (QL).

The idea that QL should be measurable would at first appear suspect. Henri Bergson ([4], Ch. 1) developed a number of celebrated arguments against the reduction of the qualitative to the quantitative, reproaching the psychophysicians for having (in order to make psychology "scientific") passed off, against all likelihood, the perception of magnitudes as a magnitude of perception. For Bergson, it is for instance illegitimate to attempt to say *by how much* one pain is more intense than another, two sensations of pain being always qualitatively different. But medical semiotics goes against Bergsonian radicalism in distinguishing qualities of pain (throbbing, searing, colicky, shooting, piercing, etc.) and intensities of pain (the throbbing of a headache can be more or less intense). Patients who are asked to evaluate the efficacy of an analgesic do not find anything absurd in saying that one analgesic reduces their pain more than another. (I shall be allowing below that well-being can have several components ("dimensions"), but shall be saying that within each component it admits of degrees.)

QL from the individual point of view, is what we wish one another at the turn of the year — not simple survival, but what makes life good: health, love, success, comfort, joys — in short, happiness. From the collective point of view, quality of life is not reducible to economic prosperity (standard of living and level of development); it involves political goods (liberty, equality, security), cultural goods (education, information, creative freedom) and demographic resources (suitable birth-rate, population globally in good health, low mortality). The notion of QL is certainly pluridimensional [8] [42]. Even, indeed, if one limits oneself to the part of quality of life that is related to health (HQL), one is confronted with a pluridimensional concept. Let us recall the WHO definition (1947): "Health is a state of complete physical, mental and social well-being and not merely the absence of disease or infirmity" [32]. With certain variations, recent works on the "profiles" of HQL retain these three dimensions: physical functioning, mental functioning, social functioning [19] [40] [64] [65] [20]. That within a dimension there are degrees in the quality of functioning is pretty easily granted — everyone understands that a disease or infirmity can either more or less seriously alter one's physical or mental capacity, for instance. There has been a greater amount of discussion concerning whether to consider deterioration as being indicated by "objective" limitations of performance or by "subjective" degrees of dissatisfaction. Doctors would tend to want to "objectively" evaluate the loss of capacity caused in their patients by a given pathology. But today there is agreement in thinking that it is more worthwhile to obtain the ("subjective") opinions of the patients themselves [7] [5] [13]. Whether it is legitimate to reduce a "structural" (pluridimensional) profile to a single (numeric) index of deterioration of HQL, and how this index (of the QL) can be synthesized with the quantity of life index, is the subject of even more debate. An elegant method for measuring the "loss" of

QL occasioned by various health hazards illustrates how such a project can be carried out. This method, proposed by a team at McMaster University (Ontario) in the 70s [61], consists of asking the subject to compare in his mind the following two possibilities: (a) living to the end of his days with a given disease, and (b) living a shorter time but without this disease. The more the disease is feared, the greater the amount of life the subject is prepared to exchange for the possibility of being spared from it. This "time trade-off" method has its place in the conceptual framework of the theory of decision-making under uncertainty [27] [28]. It furnishes a measure of the utility, or desirability, of various states of health. Experimental evaluations made with the help of several population groups or samples give, by way of indication, the following results. Taking (by convention) the figure 1 as indicating being in good health, and 0 as indicating being dead, then living under treatment (with side-effects) for arterial hypertension is rated 0.95, living with a kidney transplant is rated 0.84, living under iterative dialysis at hospital is rated 0.57, living with severe angina pectoris is rated 0.50, living paralysed and bedridden in hospital (without pain) is rated 0.33, and surviving in a state of chronic vegetative coma is rated less than 0 [62]. Other evaluations have been made with other methods, the respective merits of the various methods have been discussed, and an attempt has been made to ascertain what degree of consensus the obtained scales can command [48] [54] [33] [26] [38]. In short, during the past twenty years or so an enormous effort has been made to construct simple and reliable instruments for measuring HQL.

That a person can prefer to cut short his life rather than to survive in conditions that he considers incompatible with his dignity, is an ancient idea, attested in philosophical tradition. Socrates told his judges that he would rather die than lower himself to supplicating them ([49], 178e). The Stoic sage claimed the freedom to escape by death if captured ([16] IV, 1 (29)). That this attitude should be carried over to the collective level, in order to elaborate a health policy, is a less familiar idea, but an idea nevertheless which inevitably comes to mind in the face of certain accomplishments of modern medicine. Thus when the dentist Barney Clarke volunteered to receive a Jarvik heart, and the media coverage made it clear for all to see that this man permanently attached to his machine had a miserable life, it became common knowledge that there are "services" which medicine should not offer. Keeping sick persons alive is good, but still it is necessary that the time they have left should be tolerable. If, in addition, poor-quality survival has been obtained at exorbitant cost, the thought comes to mind that the money in question could have been better spent. There are excessive investments which one must know how to avoid. The wise management of health budgets — which are not infinitely extendible — necessitates reflection on the cost-effectiveness (or cost-benefit) of the investments that have been authorized [66]. And the effectiveness of an action performed in respect of health can be measured by the gain in quantity of life that it brings about, adjusted with respect to the quality of that life.

The notion of "quality-adjusted life years" (QALYs) has been proposed by some doctors and/or health economists for justifying medical decisions [39]. If the QALY figure for one year of life in good health is 1, the figure for one year of life in poor health is less than 1. If certain forms of survival are worse than death, their figures are negative. The envisaged result of an action in respect of health is calculated in QALYs, or in "quality-adjusted life expectancy". This represents a refinement on the calculations of Bernoulli and Duvillard. The action is effective and fruitful in proportion to the number of QALYs it engenders. It is efficient to the extent that the cost of its QALYs is low. Here again, the adjusting of a quantity of life gained (or of a probability of survival) by its quality presupposes that the quality can be translated into quantitative terms. The supporters of this type of rationality maintain that the sick persons understand very well the questions put to them in this context (and that these persons would rather have the questions put to them than have someone else decide for them). Take the case of a woman with breast cancer. The doctor says to her: "There's a choice between two types of therapy. One of them has an 80% 5-year survival rate but involves a mastectomy, meaning you've got to accept a mutilation. The other one has, so far as we know, a 78% 5-year survival rate but involves only a tumorectomy, meaning you keep your breast." Different women will choose differently, but anyway such a choice will make sense to them. Concerning choices that are no longer individual but global, it has for instance been calculated by the QALYs method that it is more worthwhile to set up renal transplantation units than dialysis units, and more worthwhile to prescribe oestrogens to post-menopausal women (to combat osteoporosis) than to have to insert hip prostheses when osteoporosis leads to fractures [67]. The rule is that in order properly to serve its beneficiaries a health system must produce the maximum QALYs at the minimum cost.

Nevertheless, though the rationality of QALYs may appear to bear the stamp of common sense in ordinary cases, it is not certain that it offers the best solution for medical dilemmas. Michael Lockwood, who makes no secret of his reservations, recalls the story of the selection committee of the

Seattle Artificial Kidney Center (familiarly known as the "God Committee"). Between 1961 and 1967, the institutions being inadequate to meet the needs, this committee was charged with choosing which persons were to be granted haemodialysis. They later published their criteria, saying that in order to select the best candidates for treatment they had taken into consideration "sex, marital status, and number of dependents, income, net worth, psychological stability, and past performance and future potential" ([6], p. 233, note 111; cited in [39], p. 318). Which is to say that a person who was elderly, unmarried, on a low socioprofessional level, had no chance. Such a person was allowed to die of renal failure. There are today in certain North American medical centers lists (published) of criteria that serve to establish an order of priority regarding persons awaiting an organ transplant. Lockwood argues that, the more explicit and detailed the qualitative criteria, the greater the risk of their confirming the prejudices of those applying them. He notes that in the UK there are very few black people undergoing dialysis. If the surgeons have just one kidney (or heart, or liver) to transplant, and there are two candidates whose lives depend on a transplant, and one of these candidates is young, white, qualified, whilst the other is old, black and unqualified, then a calculation in terms of QALYs will always indicate to these surgeons that they should choose the former. Lockwood recalls ([39], p. 313) that in 1986, at the annual assembly of the British Medical Association (BMA), a dilemma of this type caused a storm. After the economist Alan Maynard had defended QALYs, explaining that it would be irrational to put a feeble elderly man on dialysis and to let die beside him a young man with a family to look after, the philosopher John Harris retorted that the calculation of QALYs might be rational but was unfair, involving a systematic discrimination against the oldest and weakest, and said that if one of the two candidates had to be chosen it would be better to draw lots.

Here two premises confront each other. They link up with two great traditions of moral philosophy: on the one hand the teleological tradition (J. S. Mill [41], taking up again the Aristotelian issue of the highest good), and on the other hand the deontological tradition ([29], II, that refuses to derive moral law from a notion of the good).

For the teleologist the object of the moral life is to make the world as good as it can be for all sentient beings (which is to say, beings capable of suffering). The fundamental moral question is: what is the best state of affairs? The value of an act is measured by its consequences. Ethically correct conduct is that from which results the maximum of good (happiness) and/or the minimum of evil (unhappiness). Classical utilitarianism [3] postulates that global benefit is the sum of individual benefits, and calls for the maximization of benefit or "utility". If it be admitted that individual interests can enter into conflict, and if the maximization of global interest be prescribed, one has a form of utilitarianism that Amartya Sen [56] has proposed calling "welfarism", where the good of the community of sentient beings can imply a minimum of individual sacrifices. A welfarist recognizes that there are moral dilemmas, and assumes the responsibility for resolving them in terms of what is globally better (or less bad). Thus the British philosopher R. M. Hare [24] argues that the therapeutic termination of pregnancy can be morally good. A couple who cannot bring up more than one child choose to have a deformed foetus aborted in order later to have a normal child. They replace a deficient life by a life of good quality — and that is in conformity with the principle of the best (or Greatest-happiness Principle, as Mill called it).

The deontologist holds himself responsible for his own attitude, not for its consequences in respect of the state of the world. The fundamental question for him is: what is my duty? For there are things with regard to which a selfrespecting being feels an obligation (as the saying goes, *noblesse oblige*). And there are things that a self-respecting being refuses to do. The deontologist does not necessarily regard the calculation of advantages as immoral. Establishing renal transplantation units rather than dialysis units involves a technical choice — and it can be a good choice, giving a satisfactory QL to a maximum number of persons with renal failure. But the deontologist will not accept the subordination of the moral law to an empirical notion of the good ([30], I, I, 2). There is an imperative (a "categorical" one) which transcends all calculation of good or of happiness. The dignity of existence of a moral subject is not a negotiable good ([29], II). The deontologist considers for instance that no reasonable being can want to sacrifice one sick person for the sake of another. Certainly, when confronted with two persons with renal failure, it is possible on the emotional plane to wish to save the one who is young, qualified, and has a family to look after, rather than the one who is elderly, unqualified and unmarried — but wishing is not itself moral. No reasonable being can want to dispose of the life of one in order to save the other (for no end, however good it may be, can justify such a means). Nor can any reasonable being even ask one to sacrifice himself to the other, because an autonomous being cannot want his own elimination. The deontologist refuses to choose — he lets "nature" choose instead of him (he advocates the drawing of lots).

Medical ethics is wedged between these two attitudes. Their professional life constrains doctors to make "tragic" choices [59], and to make cost-effectiveness calculations to elucidate these choices. But their deontology proclaims an egalitarian and humanistic principle of "the sacred nature of human life" which Helga Kuhse [37] affirms is in contradiction to what they do. However, this type of conflict is not reserved for doctors. Utilitarian theorists, engaged in scrutinizing the rationality of QALYs, have explored the limits where it becomes evident that not everything is commensurable. And those who reject consequentialism in theory, reserving the right to "wash their hands of" what happens, take steps to ensure that in practice the consequences of their actions shall not be too bad.

Michael Lockwood thinks it legitimate in general to have recourse to a calculation of the QALYs type as a *guide* to making choices in the field of health. "Any reasonable moral theory necessarily involves, it seems to me, an element of welfarism: other things being equal, one should hold it to be morally preferable to produce a greater global benefit" ([39], p. 316). One would have to be, he says, extraordinarily sceptical not to recognize that the utilitarian effort to rationally analyze decisions is based on a lucid and bold principle and has produced solid results. He differentiates between two levels of choice, the global and the individual. A hospital manager, a regional health board, Parliament when it votes on the health budget, decide what resources (scanner, operations facilities, transplant unit) are to be available to what populations (macro-allocation). The midwife, the doctor, the nurse, decide who is to receive what specific care at what specific time (micro-allocation). On each level, effective and sensible distribution, arranged in accordance with the principle of the best, and supported both by reflection concerning the future and by precise indices of quality, is in general preferable to a random or arbitrary distribution, especially where the latter involves an element of whim or favoritism. If it is necessary to choose between setting up dialysis units and setting up transplant units, it would seem desirable to relate this choice to the preferences and interests of the beneficiary population, with a concern not only to meet the needs of today's patients but also to prepare a better situation for tomorrow's. And if there are fewer hearts to transplant than persons waiting for a cardiac transplant, it would seem desirable that the positions on the waiting-list should be determined by *public criteria giving priority to those who are best able to benefit from the transplant* (to avoid the waste of a scarce resource), rather than be determined by who you know or by money under the table, or even by the drawing of lots.

And yet, says Lockwood ([39], IV), the welfarist wisdom finds itself having to face the claims of justice. There is a point at which a rational distribution, whatever its nature, is judged inequitable.

The limits of utilitarianism have been the subject of noteworthy analyses, often carried out by the utilitarians themselves. But the first objection comes from the other side. John Rawls has defended a theory of justice which postulates that the just has priority over the good ([53], I, 6), and that a social organization should be chosen which tends permanently to correct natural inequalities in that it restores an equality of chances and of freedoms ([53], II, 13). Now, accidents to health are (typically) natural injustices which diminish our chances and our freedoms. Therefore it is expected of the health system that it shall be organized in such a way as to cure disease, compensate for handicaps, rehabilitate the person who has been afflicted, and it is further expected that the system shall devote greater resources to those who are most ill than to those who are in better health. In Rawls's perspective any difference that benefits the most disadvantaged, in restoring chances to them, is admissible. So it should not be a cause of complaint that the treatment of a person with cancer costs those fortunate enough not to have cancer a vast amount of money. The injustice would be to refuse aid to the person already weighed down with misfortune. In line with this it has been argued that the optimal distribution in accordance with the calculation of QALYs is not automatically just, that it tends to aggravate certain inequalities in favoring those persons who are already relatively fortunate (having anyway a minimal quality of life) and penalizing those who are already less fortunate (in particular the elderly). Lockwood does not entirely agree with this objection, because if one were to push it to the limit one would be confronted with the absurdity of having the largest proportion of health resources directed to the care of persons whose condition is hopeless, neglecting the less seriously ill who can be cared for more effectively. He does not find it unjust if in a situation of acute shortage an elderly person is refused dialysis, or an organ transplant, in order to give a younger person this benefit. The elderly person, says Lockwood, has already had his chance, has had his life. But except in the case of extreme situations, he thinks that the principle of equality should have priority over that of benefit. This view is related to the professional ethics that prescribes that the doctor shall divide his time and attention equally between sick persons that he can cure and those for whom he can do nothing.

The rule saying that it is necessary to "maximize" the quality/quantity of life gained raises yet another problem (seemingly a technical one): Is it the sum of gains that is to be maximized, or is it the average? This inoffensive-looking question is raised by Robert Nozick ([43], Ch. 3) in a well known passage where, enquiring as to whether there is any limit to what we can do to animals, he advances the rule: "utilitarianism for animals, Kantianism for people". This rule implies at first sight that one can sacrifice an animal, but not a human being, to improve the global situation of the world. For example, one cannot take the heart of a human being to save another human being with it. But one can sacrifice an animal, take out its heart, and transplant it into a human being (on condition perhaps that such an animal has been expressly bred for this purpose, that is, if it otherwise would not exist at all). Let us reflect now on the benefits created by a health system of strict utilitarian orthodoxy, when the utilitarian moral code applies also to people. If it is a question of maximizing the sum of benefits (as can always be done by augmenting the number of beneficiaries), it will be sufficient to provide health care of a quality just above the average, and augment the beneficiary population, to satisfy the principle of the best. That, says Nozick, is "inept". But if it is a question of raising the average, it is enough if a single human being benefits immensely from the health system, even if it should mean killing everyone else, for the average benefit to sky-rocket, imagine, say, a health system which devotes all its resources to making a single human being immortal and well-functioning, with the rest of the population serving as a source of spare organs and tissues. That is "monstrous"[2]. Reconsidering the double paradox, Derek Parfit ([45], Ch. 17; [46]) has described as a "repugnant conclusion" that to which the arithmetical application of the principle of the best leads. Given a situation in which humanity has a certain level of well-being, there follows a "better" situation either (by augmentation of the sum of well-being) when the population increases faster than individual well-being diminishes, or (by augmentation of the average) when the wellbeing of a (developed) minority increases much faster than that of the (underdeveloped) majority withers away. Now, it is found "repugnant" to admit that the number of lives compensates for their poor quality or that the extreme quality of certain lives compensates for the poor quality of the others. Neither the distribution of rudimentary care to the whole of an exploding population (by "barefoot doctors"), nor a two-tier health system that superbly takes care of a small number of privileged persons (in "reserved hospitals"), represents the "best" organization of health care.

R. M. Hare [25] says that there will be a natural correction, because the actors in a utilitarian system will spontaneously tend to modify what in the evolution of the system offends their sense of justice. That does not imply their leaving the system. In the first place, an abuse of rationality that offends people's feelings is not optimal in the utilitarian sense. Then in the second place, technical solutions can be developed in order to prevent paradoxical deviations. If one has calculated that putting in hip prostheses produces more QALYs than transplanting livers ([39], p. 320), and one is shocked at this because a liver transplant is more "vital" for the beneficiary than a hip prosthesis, one can modify the way options are rated, or reject the comparison. As a matter of fact, it suffices to refrain from reducing quality to quantity, and to preserve for quality its multi character, to escape the paradoxes of compensation. It has been argued that scientific results will be more in tune with intuition the more the indicators of well-being are characterized by nuance and diversification, and the greater the respect shown for the structure of each individual perception of QL [20]. After all, the well-being of the gardener is not that of the engineer, and the hip arthrosis which ruins the QL of the former may hardly trouble the latter. But if one pursues this logic to the limit one is confronted with the fact that each individual existence is qualitatively unique, incomparable and irreducible to all the others, so that finally the notion of the "best" choice becomes devoid of meaning. What constitutes the value of my life belongs to me alone. There are no lives better or less good than others, there are just different lives, all of which enrich the palette of humanity, and which there is no reason to standardize or optimize, even by way of sanitary generalizations.

Bernard Williams [68], with mordant scepticism, has analysed the "slippery slope" type of progression. A teleological ethics always has trouble with extreme situations, where its optimisation principles — which produce fine results in average cases — may well yield results falling far from the "happy mean". We have seen (above) the slope leading towards "repugnant" extremes. Here, in order to avoid slipping towards the "arbitrary result" or the "horrible result", the welfarist now slips towards the very thing he wanted to avoid, i.e., holding the QL to be an absolute which defies all comparison.

The representatives of the opposition celebrate a momentary triumph: the inanity of the QALYs type of calculation is recognized even by its supporters! Such a calculation is repugnant to deontologists, who are horrified at the idea that the life of a 95-year-old with renal failure should be less precious than that of a 40-year-old suffering from the same affliction, or the idea that the abortion of a fetus afflicted with trisomy should raise fewer moral objections than that of a normal fetus of the same age — in short, deontologists are horrified at the idea that all lives should not have the same value. They draw attention to the danger of making us, under the cover of medical science, the arbiters of which lives deserve to be lived. They affirm that every life has, in itself, an absolute value, and that it is immoral on the part of a human collectivity to have its members believe that health expenses are an investment which in their case is perhaps not worthwhile. Everyone is worthwhile in this respect, every life is as "sacred" or "inviolable" as any other, every form of life merits respect.

Let us here clear away an ambiguity. The adage that "Life's worth nothing, but nothing's worth life" indicates that life (my life) is the *sine qua non* for anything at all to be worth anything (to me). Without (conscious) life, nothing is worth anything. Life does not have value, it is the condition for values. Or rather, its value is incommensurable with other values. But on the other hand we commonly admit that there are goods (political, economic, aesthetic) for which it is worth giving one's life, for which it is worth giving up some of one's life-time. Furthermore modern human communities know very well how to calculate the cost in human lives of the extraction of off-shore oil, of motor traffic, or of a war undertaken to regain some territory — the price of human life is then perfectly commensurable with that of other goods. Those who advocate a deontological ethics are not unaware of this fact. They can escape the problem only by supposing that no one is forced to drive a car or work on an oil-rig, and that military service is freely assented to within the framework of the social contract. It is this (transcendental) freedom of the individual which is not negotiable, not his life in the empirical sense. This distinction of levels is crucial.

There are two philosophical traditions regarding respect for life, both of which have religious roots. In the oriental tradition (Buddhism, Jainism) — to follow the exegesis of it provided by Schopenhauer — phenomenal (empirical, individual) life is at the same time both suffering and illusion, ephemeral and unstable, squandered by nature, spared by the compassionate wise man. That which is immanent in life (the being-in-itself, the universal will to live) is "indestructible" and "imperishable" [55]; it is foolish to say that it shall not be killed — it cannot be (metempsychosis). In the Judaeo-Christian tradition the commandment "Thou shall not kill" is absolute, not because of the intrinsic value of human life (for indeed human life has very little value, being "a mere breath", "a shadow", as Psalm 39 has it) but because of the transcendent and sacred nature of the divine will. To do away with a life is to insult its creator, who had judged that it deserved to be. If one follows the lay and philosophical version of Judaeo-Christianity provided by Kant, what is deserving of absolute respect in the universe is the existence of free and reasonable creatures, capable of affirming what should be (universally) and of undertaking to do what they should (irrespective of their inclinations) ([29], II). And what should they do? Respect in themselves and in others that freedom which constitutes their dignity. When it comes to medical practice, this means that every sick person shall in the first place be supposed autonomous, which in turn means that decisions concerning him shall not be taken in his place when he is capable of taking them himself ([15], Ch. 7). And when it comes to health policy it means that nothing will be done for people's own "good" in spite of them, and that broad lines of policy shall be submitted to democratic control.

That does not get rid of the dilemmas. As mentioned before, the deontologists have a horror of resolving them when that means that the will or the freedom of others (or that of God, for the believers) is made subservient to an idea of the good. But they have as great an aversion as others to the drawing of lots. And to do nothing, or to let "nature" take its course, is often equivalent in practice to deciding where the "good" is to be found. What is more, the principle of non-intervention in the decisions of beings presumed autonomous does not exclude a paternalistic or directive attitude with regard to beings presumed bereft of autonomy. And when one decides for another person, does one not act "in his interest" [36], or "for the best"?

The slipping of the deontologists towards an ethics of beneficence, or towards a Thomistic type of synthesis between an ethics of right intention and a consequentialist ethics, can take at least three directions: redefinition of the limits within which the deontological principle is applicable, indulgence for the sinner, and handling of cases of conscience (or dilemmas) by means of a casuistry based on the principle of the best.

For the past forty years there has been much discussion, especially in North America, concerning the necessity said to exist for "redefining death" ([18], A01, pp. 17-31). Such discussion started in connection with the first attempts at transplanting vital organs (renal transplant, 1952; cardiac transplant, 1967). The removal of an organ from the "corpse" presupposed, for the organ to be living, that the corpse was in a correct haemodynamic state. In the United States the first cardiac transplants led to legal gaucherie: the organ was removed, then the donor declared dead — in conformity with the sole criterion of death acknowledged at that time by American law, the arrest of cardio-respiratory functions. The efforts of several committees of experts, including that of Harvard [2], made it possible to set forth clearly, and to have acknowledged by the law, that the arrest of cerebral functions was a criterion of death as valid as the preceding one. Thus the proper procedure was to declare the donor dead in accordance with the criterion of the arrest of cerebral functions, at the same time artificially maintaining his cardio-respiratory functions, then to remove the organ. Despite the fact that it was sometimes thought to be so, it was not in fact a question of a new "definition", because the arrest of all cerebral functions diagnosed by the Harvard criteria, causes cardio-respiratory arrest, and the other way round. Death remains what it has always been: the irreversible cessation of functioning of the organism as individual totality. But the case of persons surviving for years in a state of chronic vegetative coma, with subcortical cerebral functions more or less intact but with cortical functions gone, led some to wonder whether it would not be appropriate to declare these persons dead, and to stop medical care, when there was no hope of recovery. The American President's Commission [51] was tempted to define human death as the definite cessation or absence of all relational life ("cortical death"), but shrank from doing so. It was a question of a real "redefinition", which would have made it possible to declare the death of organisms capable of surviving for a time on the biological plane but incapable of a properly human life, for instance anencephalic new-born babies or the "vegetables" in a state of vegetative coma. The Law Reform Commission of Canada went further, saying that the redefinition was justified but was premature "in the present state of society and of medicine" ([10], p. 17). Reproaches were levelled at the American Commission for being timorous — principally by neo-Kantian moral philosophers [69]. For a utilitarian the concern with definition is unimportant: there is nothing sacred about human life, and there is nothing shameful about openly posing the question of euthanasia when this life loses its quality to the point where death seems better. A deontologist, by contrast, shrinks from the question of euthanasia. For him the person of the other is inviolable. A person in a coma may always recover, even though the probability of recovery may be infinitesimal. Nevertheless, there are cases where this probability is zero. In order to be able to at the same time handle these cases and preserve the absolute principle that "Thou shalt not dispose of the life of thy fellow-being", it is very important to be able to say: here there is no longer a fellow-being, there is not really a human person, there is a life without human "quality", there is nothing but a vegetable. Similarly, the utilitarian unconstrainedly discusses the good or bad indications regarding the termination of pregnancy, whilst the deontologist tends to regard it as impermissible as a matter of principle, though sees that in certain cases not to permit it is worse, and escapes from this by positing a threshold [17] beyond which it is so evident that the human embryo is not an autonomous being that the principle is inapplicable.

A second way of firmly maintaining an absolute moral imperative whilst at the same time making concessions to the utilitarian calculation is to indulgently shut one's eyes to human weakness. A pregnant woman learns that the child she is expecting is afflicted with major thalassaemia. If "out of duty" she chooses to bring the child into the world and to accompany it in the few years of sick life that it has before it, she will be held in esteem. The deontologist cannot force her into it. He assumes that she is reasonable and autonomous; it is up to her to choose. But being assumed reasonable, she cannot want to universalise the rule that abortion is permissible, because no rational human being can want the possible self-destruction of the human race; nor can she want to universalise the rule that selective abortion for thalassaemia is permissible, because that would amount to deciding who is worthy of living. Choosing to terminate the development of a thalassaemic fetus is not choosing to do one's duty, it is choosing to avoid unnecessary suffering. This choice is humanly comprehensible, because human beings are weak. Though not approving the choice, and not according it moral value, one can shut one's eyes to it and refrain from censuring it. It is even possible to officially condemn the termination of pregnancy in general, yet tacitly favour the eugenic termination of pregnancy in the case of thalassaemia, as does the Catholic Church in the Mediterranean basin. Double-dealing? One way of coping with the problem of the rule and the exception. Where the teleologist would tend to want to make the exception compatible with the rule, by subtly modifying the rule, the deontologist prefers to keep a strict rule, and tolerates the exception without comment. Then finding that certain exceptions are more acceptable than others becomes the occasion for nuances of humanitarian sentiment rather than of ethical rationality. Thus in countries with a strong deontological tradition there exists a resistance to the idea that a law can clearly define, even in a very restrictive fashion, in what cases a doctor can have recourse to euthanasia for the benefit of a sick person. It is preferred to retain the principle that euthanasia is always forbidden, whilst knowing that it is in fact practised on a compassionate basis. This attitude goes hand in hand with a certain moral elitism: for the majority the rule in all its rigour, for an enlightened minority the privilege of discerning when the rule may not apply.

It does happen, however, that the deontologist indulges in casuistry, which is to say he reasons concerning particular cases in order to specify the circumstances in which exceptions to the general rule are admissible. Thus the Catholic Church has on several occasions issued prudent directives concerning the care of the dying [63]. These directives incorporate considerations of quality in two ways: by accepting the use of sedation even when this involves a risk to life, and by permitting the judging of the moment when one is no longer acting for the good of the sick person in inflicting upon him troublesome means of resuscitation in order to keep him alive. It is a question of clearly defining the cases in which euthanasia is morally legitimate. Two types of argument are used: one type rests on the distinction between nature and artifice, the other on the principle of the double effect. Pius XII (1957), distinguishing between "ordinary" and "extraordinary" means of resuscitation, then the Congregation for the Doctrine of the Faith (1980), distinguishing between "proportionate" and "disproportionate" means, say that it is necessary to know how to let a sick person die, when one becomes aware that he is in any case near death, rather than to attempt to resist this death by artificial means. There has been much discussion concerning the classification of the feeding and hydration by catheter of a patient in a persistent state of vegetative coma: is it to be classified as ordinary care, or as "artificial" medical procedure? The American Academy of Neurology declared in 1988 that it is an artificial procedure [1]. If so, to shorten life by suspending this procedure is legitimate. By the same argument, it is legitimate not to submit to dialysis in the case of renal failure the elderly and infirm person suffering from dementia. This argument is not generalizable unless it includes an assessment of the quality of the life thus prolonged or shortened — otherwise it would permit refusing a lifesaving antibiotic to a child who is dying of an infectious disease.

The arguments based on the distinction between the effect intended and the effect produced make it possible — without having a consequentialist ethics — to obliquely calculate the acceptability of the consequences of what one does. Thomas Aquinas said ([60], II, 1, Q20, A5) that the consequence of an act, if it is different from the end intended by the will of the agent, can modify the moral value of the act. But he added that if the consequence is not foreseen with certainty, nor really foreseeable, the value of the act is not affected by it. Thus it will be argued that the use of a sedative to quell the pain of a sick person (first effect, intended) is legitimate, even if it involves a risk of precipitating death (second effect, consequence), on condition that the dosage be such that death is not certain. Sarcastically, Pascal affirmed that "this method gives you the right to kill... provided you direct your intention properly" ([47], p. 733) (that is, provided you do not intend it directly, but only obliquely, as a second effect). Kuhse, having thoroughly examined how the principle of the double effect is put into practice in the medical context ([37], Ch. 3), states that it serves to disengage the doctor from responsibility. The doctor treating a seriously ill pregnant woman with a drug capable of bringing about a miscarriage does not judge himself directly responsible for the miscarriage if it should occur, even if the treatment is indeed the cause of it. The doctor who omits to vaccinate against influenza a person who is elderly, infirm and suffering from dementia, on the pretext of sparing the person trouble, does not regard himself as having killed the patient if the latter dies of influenza, even if the thought has crossed his mind that it would be a blessing for the person to be carried off by the infection. Kuhse, to tell the truth, does not understand these casuistic subtleties. Convinced utilitarian that she is, she believes that if the doctor concludes that death is a good in a given case, then he should be able to assume responsibility for bringing it about, and that he is no less responsible for it if he does nothing to prevent it than if he directly induces it. From the point of view of an intentionalist ethics there is on the contrary all the difference in the world between assenting to the death of another person and wanting to kill that person. It is certainly no accident that in countries where the deontological tradition is dominant, and utilitarianism held in contempt, there is an antipathy to having a free and open debate on health policy. It is indeed very difficult to consciously assume responsibility for the human consequences of the economies that are indispensable in order for the system of care to remain globally effective. It feels as if one wanted people to die if one actively decides to restrict the access to certain expensive treatments.

When there coexist, in a democratic society, persons with different philosophies and moral sensibilities, the procedural solutions are those that prevail if it is necessary to define a rule that shall apply to all. If in the case of euthanasia, for example, certain persons demand that there be written into the law the exact conditions under which it is permitted, whilst others want the law to still penalise it in principle, albeit shutting their eyes to certain infractions, the only point on which they can agree is that there needs to be a public debate. The *a priori* of communication [23] is the lowest common ethical denominator for those who accept living together. "Let's discuss it: explain to me why you want things to be like that, and I'll explain to you why I want things to be like this, then at best our positions will converge, but if we still can't agree we'll settle it by a vote."

But what comes from the discursive exchange between advocates of a teleological ethics and advocates of a deontological ethics? (What degree of fatality is there in confrontations like that of Wichita?)

Very few people have an articulated and coherent moral theory, few are clear about why they think that things should be so rather than otherwise. Our common morality is a mixture of conformism, fine feelings, and scraps of heterogeneous traditions: Stoic, Cynic, Christian, etc. Medical ethics itself is a mixture of manifestly teleological precepts (firstly not to injure, then to do everything to improve the health of the sick person, and finally to contribute to the improvement of public health) and of deontological precepts (to respect private life, not to practise discrimination). Is it because of this theoretical blur, or in spite of it, or through the confrontation with the specific situation, or through a regulatory effect of the infra-discursive communication itself, that it is in general quite easy to reach a consensus regarding a concrete case? When it comes to a decision about abandoning or refraining from attempts at resuscitation, there is often little to "discuss". If there is good communication between the medical team, the family and (when he is competent) the sick person himself, the "practice policy" will gradually reveal itself, without the underlying ethical (or religious) convictions having to be made specially explicit. If a real ethical uncertainty and/or serious divergence of opinion comes to light, as has happened in North America in several cases of persistent vegetative coma, there are procedures available: discussion of the case by hospital ethics committee, or arbitration by tribunal [44]. During such a procedure the arguments need to be honed, but (even in North America) these individual cases where argumentation becomes necessary are an inconsiderable minority.

Things are different when it is necessary to determine a collective rule of conduct or a health policy. Hazy consensuses are fated to evaporate, whether they assume unanimous adhesion to principles that do not in fact produce unanimity, or whether they mix contradictory imperatives between which there has been no wish to choose. Communicational wisdom (if such there be) consists, not in seeking a consensus, but in not being afraid to see the validity of the opposing position ("Listen to your adversaries first, and the rest, perhaps a consensus, shall be added unto you.").

It is not an easy or spontaneous wisdom. In the mouth of a deontologist the epithet "utilitarian" is often pejorative, if not quite simply synonymous with "immoral". Kant himself considered that to subordinate free will to a principle of the good, which is to say to base a system of morals on the idea of happiness or perfection (even if it be divine perfection) is "absolutely opposed to morality" ([29], II). The other way round, the teleologist is shocked by the rigidity of the theories of duty, by the divergence they tolerate between rule and practice, and by their apparent indifference to considerations of welfare, or of improvement in welfare. Such misgivings can be overcome, though this does not imply that there is a synthesis to be found at the end of the discursive path. We have seen, certainly, that each theory has a propensity to slide towards the other. It needs also to be recognised that each is capable of incorporating elements of the other. Since Mill ("better to be a human being dissatisfied than a pig satisfied": ([41], Ch. 2), all utilitarians have said that for a competent human being the most important component of quality of life is the exercise of his autonomy; not suffering comes after. And in the case of those deontologists who accept the idea that man has a responsibility in the history of the world ([9], Ch. II, p. 3), a principle such as that of the "sacred" nature of life becomes open to being put to the test in real situations, and coexists with the principle of the best. However, the utilitarian looks for empirical universals, whilst the deontologist remains attached to formal universality. And the convergence of the two approaches is by no means guaranteed, nor the superiority of one to the other. Is it the respect for individual free-will, or the concern for the public welfare, which is to have the last word when it comes to morality? Is a soundly welfarist health policy compatible with a medical ethics that bears the imprint of personalist humanism? The wager of what is called ethics of communication is to let each person state an opinion on the matter, and argue as strongly as he can, without slipping into scepticism.

NOTES

[1] Original publication: Fagot-Largeault, A. (1991), 'Réflexions sur la notion de qualité de la vie', *Archives de philosophie du droit*, Volume 'Droit et science', 36: 135-153. English translation by Malcolm Forbes.

[2] "Utilitarianism is notoriously inept with decisions where the number of persons is at issue. (In this area, it must be conceded, eptness is hard to come by.) Maximizing the total happiness requires continuing to add persons so long as their net utility is positive and is sufficient to counterbalance the loss in utility their presence in the world causes others. Maximizing the

average utility allows a person to kill everyone else if that would make him ecstatic, and so happier than average" ([43], Part I, Chap. 3, End of Section 'Constraints and animals').

BIBLIOGRAPHY

[1]   American Academy of Neurology: 1989, 'Position of the AAN on certain aspects of the care and management of the persistent vegetative state patient; and Commentary', *Neurology, 39*, 123- 126.

[2]   Beecher, H.K.: 1968, 'A definition of irreversible coma. Report of the Ad Hoc Committee of the Harvard Medical School to examine the definition of brain death', *Journal of the American Medical Association, 205*, 337-340.

[3]   Bentham, J.: 1789, *Introduction to the Principles of Morals and Legislation*, Oxford.

[4]   Bergson, H.: 1889, *Essai sur les données immédiates de la conscience*, Alcan, Alcan.

[5]   Bernheim, J.L. and Buyse, M.: 1983, 'The anamnestic comparative self-assessment for measuring the subjective quality of life of cancer patients', *Journal of Psychological Society, 14*, 25-38.

[6]   Calabresi, G. and Bobbit, P.: 1978, *Tragic Choices*, Norton, New York

[7]   Campbell, A.: 1976, 'Subjective measures of well-being', *The American Psychologist, 31*, 117-124.

[8]   Cohen, C.: 1982, 'On the quality of life: some philosophical reflections', *Circulation, 65* (suppl 3), 29-33.

[9]   Commission de Réforme du Droit du Canada: 1979, *Le caractère sacré de la vie ou la qualité de la vie*, Rapport établi par E.W. Keyserlingk, Série: Protection de la vie. Document d'étude. Canada: Ministère des approvisionnements et services.

[10]   Law Reform Commission of Canada: 1981, Report 15, *Criteria for the Determination of Death*, Minister of Supply and Services, Ottawa

[11]   Croog, S.H., *et al.*: 1986, 'The effects of antihypertensive therapy on the quality of life', *New England Journal of Medicine, 314*, 1657-1664.

[12]   Dehan, M.: 1988, 'Réflexions sur les problèmes d'éthique en réanimation néonatalogique et pédiatrique', *La Presse médicale, 17*(11), 507.

[13]   Detsky, A.S., McLaughlin, J.R., Abrams, H.B. *et al.*: 1986, 'Quality of life of patients on longterm total parenteral nutrition at home', *Journal of General Internal Medicine, 1*, 26-33.

[14]   Eddy David, M.: 1990, 'Practices policies: where do they come from?', *Journal of the American Medical Association, 263*, 1265-1275.

[15]   Engelhardt Jr., H.T.: 1986, *The Foundations of Bioethics*, Oxford University Press, Oxford.

[16]   Epictetus: 1956-59, *The Discourses as Reported by Arrian: The Manual and Fragment*, Loeb Classical Library, London.

[17]   Fagot-Largeault, A. and Delaisi de Parseval, G.: 1987, 'Les droits de l'embryon (foetus) humain, et la notion de personne humaine potentielle', *Revue de métaphysique et de morale*, *3*, 361-385. Repr. with modif. in: *Esprit*, 1989, *6*, 86-120.

[18]   Fagot-Largeault, A.: 1989, *Les causes de la mort. Histoire naturelle et facteurs de risque*, Vrin, Paris.

[19]   Flanagan, J.C.: 1982, 'Measurement of quality of life: current state of the art', *Archives of Physical Medicine and Rehabilitation, 63*, 56-59.

[20]   Gérin, P., Dazord, A., Boissel, J.-P., and Hanauer, M.-Th.: 1990, 'Assessment of quality of life in therapeutic trials', in G. Strauch, and J.-M. Husson, (eds.), *Recent Trends in Clinical Pharmacology. Colloque INSERM Number 186*, 143-163.

[21]   Goffi, J.-Y.: 1989, 'Argumentation éthique et justification démocratique à l'épreuve des avancées technologiques: le cas des enfants mal nés', dittoed, personal communication.

[22]   *Guidelines on the Termination of Life-sustaining Treatment and the Care of the Dying*: 1987, The Hastings Center, Briarcliff Manor, NY.

[23]   Habermas, J.: 1983, *Moralbewusstsein und Kommunikatives Handeln*, Suhrkamp, Frankfurt.

[24]   Hare, R.M.: 1975, 'Abortion and the golden rule', *Philosophy and Public Affairs, 4*, 201-222.

[25]   Hare, R.M.: 1981, *Moral Thinking*, Clarendon Press, Oxford.

[26]   Heuse, A.: 1983, 'Naissance et évolution du concept d'évaluation de la qualité', *L'Hôpital Belge, 162*, 10-25.

[27]   Holloway, C.A.: 1979, *Decision Making Under Uncertainty: Models and Choices*, Prentice Hall, Englewood Cliffs, NJ.

[28]   Jones, M.B.: 1977, 'Health status indexes: the trade-off between quantity and quality of life', *Socio-Economic Planning Sciences, 11*, 301-305.

[29]   Kant, I.: 1989, *Grundlegung zur Metaphysik der Sitten*, Klostermann, Frankfurt am Main.

[30]   Kant, I.: 1990, *Kritik der praktischen Vernunft*, Meiner, Hamburg.

[31]   Katz, J.: 1984, *The Silent World of Doctor and Patient*, Free Press, New York.

[32]   Katz, S.: 1987, 'The science of quality of life' (Editorial), *Journal of Chronic Diseases, 40* (6), 459-463.

[33]   Kind, P., Rosser, R., Williams, A.: 1982, 'Valuation of quality of life: some psychological evidence', in M.W. Jones-Lee (ed.), *The Value of Life and Safety*, Amsterdam, North Holland.

[34]   Kuhse, H. and Singer, P.: 1985, *Should the Baby Live? The Problem of Handicapped Infants*, Oxford University Press, Oxford.

[35]   Kuhse, H. and Singer, P.: 1988, 'Age and the allocation of medical resources', *The Journal of Medicine and Philosophy, 13*, 101-116.

[36]   Kuhse, H.: 1985, 'Interests', *Journal of Medical Ethics, 11*, 146-149.

[37]   Kuhse, H.: 1987, *The Sanctity-of Life Doctrine in Medicine: A Critique*, Clarendon Press, Oxford.

[38]   Leplège, A.: 1991, *Epidémiologie et décision médicale. Aspects épistémologiques et éthiques*, doctoral dissertation in philosophy, University of Paris-X, Nanterre.

[39]   Lockwood, M.: 1987, 'Qualité de la vie et affectation des ressources', *Revue de Métaphysique et de Morale, 3*, 307-328.

[40]   McCullough, L.B.: 1984, 'Concept of the quality of life: a conceptual analysis', in N.K. Wenger, M.E. Mattson, C.D. Furberg, J. Elinson (eds.), *Assessment of Quality of Life in Clinical Trials of Cardiovascular Therapies*, MTP Press, London.

[41]   Mill, J.S.: 1861, *Utilitarianism*, reprinted from Frazers Magazine, 1863.

[42]   Miller, L., Dalton, M., Vestal, R., Perkins, J.G., Lyon, G.: 1989, 'Quality of life', in: 'Methodological and regulatory/scientific aspects', *Journal of Clinical Research and Drug Development, 3*, 117-128.

[43]   Nozick, R.: 1974, *Anarchy, State and Utopia*, Blackwell, Oxford.

[44]   Orentiicher, D.: 1989, Cruzan V. Director of Missouri Department of Health: 'An ethical and legal perspective', *Journal of the American Medical Association, 20*, 2928-2930.

[45]   Parfit, D.: 1984, *Reasons and Persons*, Clarendon Press, Oxford. Repr. with corrections, 1989.

[46]   Parfit, D.: 1986, 'Overpopulation and the quality of life', in P. Singer (ed.), *Applied Ethics*, Oxford University Press, Oxford, pp. 145-164.

[47]   Pascal, B.: 1960, *Les Provinciales*, in *Oeuvres*, NRF, Bibliothèque de la Pléiade, Paris.

[48]   Patrick, D.L., Bush, J.W., Chen, M.M.: 1973, 'Methods for measuring levels of well-being for a health status index', *Health Services Research, 8*, 228-245.

[49]   Plato: 1969, *The Apology of Socrates*, Adam, A.M. (ed.), Cambridge University Press, Cambridge.

[50]   Pollini, J. and Teissier, M.: 1990, 'Un dilemme difficile à résoudre: les malades âgés récusés pour le traitement par hémodialyse itérative. Problèmes éthiques ou choix médical?', *Néphrologie, 11* (Suppl), 67-73.

[51]   President's Commission for the Study of Ethical Problems in Medicine and Biomedical and Behavioral Research: 1981, *Defining Death*, US Government Printing Office, Washington DC.

[52]   President's Commission for the Study of Ethical Problems in Medicine and Biomedical and Behavioral Research: 1983, *Deciding to Forego Life-Sustaining*

Tratment. Ethical, Medical and Legal Issues in Treatment Decisions, US Govt Printing Office, Washington DC.

[53]   Rawls, J.: 1971, *A Theory of Justice*, Harvard University Press, Cambridge, Massachusetts.

[54]   Rosser, Kind, P.: 1978, 'A scale of valuations of states of illness: is there a social consensus?', *International Journal of Epidemiology, 7*, 247-358.

[55]   Schopenhauer, A.: 1859, *Die Welt als Wille und Vorstellung*, 3rd edition, Suppl. Book IV, Chap. 41.

[56]   Sen, A.: 1979, 'Utilitarianism and welfarism', *Journal of Philosophy, 76* (9), 464-489.

[57]   Shaw, Bernard: 1911, *The doctor's dilemma*, Constable cop, London.

[58]   Singer, P.: 1987, 'A report from Australia: Which babies are too expensive to treat?', *Bioethics, 1* (3), 275-283.

[59]   'The tragic choice: termination of care for patients in a permanent vegetative state', *New York University Law Review, 51*, 285-310.

[60]   Thomas Aquinas: 1981, *Summa Theologica*, Sheed & Ward, London.

[61]   Torrance, G.W.: 1976, 'Health status index models: a unified mathematical view', *Management Science, 22*, 990-1001.

[62]   Torrance, G.W.: 1987, 'Utility approach to measuring health-related quality of life', *Journal of Chronic Diseases, 40* (6), 593-600.

[63]   Verspieren, P. (ed.): 1987, *Biologie, éthique et médecine; textes du magistère catholique*, Le Centurion, Paris.

[64]   Walker, S.R. and Rosser, R.M. (eds.): 1987, *Quality of Life: Assessment and Application*, MTP Press, Lancaster.

[65]   Ware, J.E.: 1987, 'Standards for validating health measures: definition and content', *Journal of Chronic Diseases, 40*, 473-480.

[66]   Weinstein, M.C. and Stason, W.B.: 1977, 'Foundations of cost-effectiveness analysis for health and medical practices', *New England Journal of Medicine, 296*, 716-721.

[67]   Weinstein, M.C.: 1980, 'Estrogen use in postmenopausal women — costs, risks and benefits', *New England Journal of Medicine, 303*, 308-316.

[68]   Williams, B.: 1985, 'Which slopes are slippery', in M. Lockwood, (ed.), *Moral Dilemmas in Modern Medicine*, Oxford University Press, Oxford.

[69]   Zaner, R.M. (ed.): 1988, *Death: Beyond Whole-brain Criteria*, Kluwer, Dordrecht.

PETER SANDØE AND KLEMENS KAPPEL

CHANGING PREFERENCES: CONCEPTUAL PROBLEMS IN COMPARING HEALTH-RELATED QUALITY OF LIFE

The aim of measuring quality of life will always be comparative. We may want to find out how a patient's quality of life is affected by one kind of treatment as compared with another kind of treatment. Or, we may want to find out how much quality of life is gained by investing scarce health care resources in treating one kind of patient as compared with treating another kind of patient.

'Quality of life' may be defined in such a way that such comparisons are quite straightforward. For example, it may be stipulated that quality of life has a number of dimensions, such as functional status, physical symptoms, psychological problems, etc. For each dimension a number of states may be described. Each state may be assigned a number. The quality of a patient's life may then be defined as the sum of the numbers that he scores in each dimension. Finally, by means of for example a questionnaire the definition may be developed into a measuring tool. With this tool we may be able to measure and compare the quality of life of any patient or group of patients in whom we are interested.

A measure of the sort just outlined may be quite reliable and may give results that are valid in the sense that they are similar to the results produced by other quality of life measures. There is, however, one important question to be raised about this very common way of comparing the quality of life. That is: "How do we know that the measured differences in 'quality of life' reflect differences in quality of life?"

As indicated by the question, we think that "quality of life" is not just an empty term that researchers may define as they please. In ordinary parlance the expression 'quality of life' has a number of relatively distinct meanings, and when a researcher defines the term he must indicate which meaning he intends to capture in his definition.

In this paper we shall be concerned with only one specific meaning of the term 'quality of life', a meaning which may be delineated by means of the expressions 'well-being', 'happiness' and 'subjective quality of life'. 'Quality of life' according to this meaning of the term is about what a

161

L. Nordenfelt (ed.), Concepts and Measurement of Quality of Life in Health Care, 161–180.
© 1994 Kluwer Academic Publishers. Printed in the Netherlands.

person's life is worth to the person himself. For the sake of convenience we shall refer to this meaning as 'well-being'.

We can then rephrase our question in the following manner: "How do we know that the measured differences in quality of life reflect differences in well-being?" There are two ways in which the researcher may deal with this question.

The first is by saying that he is not interested in well-being, and that he therefore does not feel obliged to answer the question. In this case we have nothing to add apart from the remark that the researcher ought to make it quite clear that he is not trying to measure well-being — contrary to what the term 'quality of life' may suggest to many people.

The second way in which the researcher may tackle the question is by saying that he is indeed trying to measure well-being, and that he wants to improve his measuring tool so that it will serve to register differences in well-being.

In this paper we shall discuss some fundamental conceptual problems in comparing the well-being of a person at different stages of a course of disease and in comparing the well-being of different persons. Our conclusion will be that such comparisons only make sense on the basis of very specific assumptions.

This conclusion runs counter both to influential philosophical theories and to what many researchers think they are doing when they measure health-related quality of life. According to these theories and according to common sense, we do at least have a rough idea of what it means to compare the pleasures and pains experienced by different people even though they take pleasure in and get unpleasant experiences from quite different things. We shall therefore begin the discussion by presenting a philosophical theory of well-being — hedonism — which, if it were true, would at least in principle allow us to compare well-being between different people.

We shall, however, argue that hedonism is an inadequate theory, and that even the kind of theory which emerges from a criticism of hedonism will only rarely allow us to compare well-being between different stages of a person's life and between different persons.

On our way we shall consider a method by means of which researchers typically translate their data so that they seem to end up with comparable "magnitudes" of well-being. However, if our arguments are sound, there is no theoretical counterpart to these magnitudes. Also, this kind of method presupposes that it is possible to rank one's well-being in a situation where one's preferences have changed. We are going to argue that this is not possible.

Furthermore, we shall discuss an attempt to solve some of the problems relating to comparisons of well-being. Our conclusion will be that at present there is no workable solution to the problems, and that therefore it is not possible to make generalizable meaningful comparisons of well-being.

This does not mean that no interesting comparisons can be made between things of relevance to health-related well-being. For example, it is possible to compare things which matter to everyone's well-being, such as the ability to walk around, be pain free and avoid depression in one's social relations. However, in many cases there is no way to compare how much they matter.

THE LIMITS OF HEDONISM

We shall take hedonism to be the theory that in all well-being there is an experienced common denominator which, when it has a positive value, may be called 'pleasure', and when it has a negative value may be called 'pain'.

The English philosopher Jeremy Bentham (1748-1832) who may be called the founder of modern hedonism gives the following description of how well-being may be measured:

> To a person considered *by himself,* the value of a pleasure or pain considered *by itself,* will be greater or less according to the ... following circumstances:
>
> 1. Its *intensity.*
> 2. Its *duration.* ... ([1], p. 64).

Bentham does not claim that all the experiences which contribute to a person's well-being are of the same type. He does, however, claim that in so far as experiences contribute to a person's well-being they share a certain quality, the quality of being pleasant or painful, and that on the basis of this shared quality we can compare the amount of well-being gained or lost in various situations. Bentham himself was primarily interested in the consequences of legislation, but if his "hedonic calculus" works it may, of course, be used also to measure the outcomes of different health-care initiatives.

If, for example, it is considered how much is gained by administering a certain pain-killer, we will have first to consider the untreated pain. How intense is it? How long is it going to last? Then we will have to consider how the patient will feel if he is given the pain-killer. Will the pain disappear completely, and if not, how intense will it be? How long will the relief last? Will there be other changes in the mental states of the patient? If so, how much pain and pleasure will be involved? Finally, we will of course also have to consider all long-term effects on the patient's mental states.

In a similar way it will be possible to measure and compare the well-being achieved by treating different patients. We may, for example, in this way find out that much more well-being is produced by giving the pain-killer to a patient with severe pain than by giving it to a patient who experiences only moderate pain.

According to hedonism, a similar calculation may be carried through for all kinds of health-care initiatives, i.e., seek to find out which mental states of the pleasurable sort such an initiative brings about. For each state we multiply intensity with duration; then we add all the different results. We do the same for all the painful states. Finally, the net "amount" of well-being achieved by the initiative is found by subtracting the pain from the pleasure.

It is also possible to compare health-care initiatives which serve to prolong the patient's life (e.g., heart transplants) with treatments which serve "only" to improve the quality of the patients life (e.g., hip replacements). The well-being in a person's life is defined by the occurrence of certain mental states. Assuming that being dead implies the lack of any mental activity whatsoever, dying therefore means a drop (or a rise) of a person's well-being to zero. The value of the extra life-years achieved by the life-saving treatments will simply be the amount of pleasure minus pain in these extra years.

Assuming the truth of hedonism we can at least in principle measure and compare well-being between different people. There are, of course, still important epistemological and methodological problems which will have to be dealt with: How do I know that other people enjoy mental states similar to my own? How do I decide which pleasurable and painful experiences a person has? How do I compare their intensity? Does it make sense to say that one experience is twice as intense as another, or does it make sense only to rank experiences as more or less intense? Attempts have been made to answer these questions [10]. We shall, however not pursue the questions, since we are going to argue that hedonism is flawed at a more fundamental level.

Hedonism as here defined rests on an assumption which according to our view (and the views of many other philosophers) is not possible to defend. This is the assumption that there is one kind of experienced quality shared by all mental states which contribute positively or negatively to a person's well-being. We shall call this "the assumption of a shared mental quality."

To see how problematic the assumption of a shared mental quality is, try to compare the following positive mental states: the enjoyment of a good meal; the pleasure of being creative; the pleasure of reading a good novel; finding out that one is not HIV-positive. It does not seem possible to discern a shared experienced quality in all these mental states. What they seem to have in common is that they satisfy us one way or the other; but that is not itself an experienced quality.

The lack of a shared mental quality should become even more obvious when the following examples of "negative" experiences are compared: nausea; the experience of intense physical pain; the anxiety when one is waiting to have a suspicion of cancer confirmed or disconfirmed; the frustration felt when one is not able to walk around or do other of those things which are normally taken for granted. The only common denominator here is that we experience something which we don't like to experience. To say that we don't like the experiences because they share a certain experienced quality is to put the cart before the horse.

Hedonism tries to explain why we prefer some things to others by saying that the preferred things give rise to a specific sort of experience. Rather, it seems that all the preferred things have in common is that they are being preferred. This, according to our view, is the main and compelling reason for giving up hedonism in favor of a theory according to which well-being is characterized in terms of preference-satisfaction. Such a theory we shall present in the following section.

WELL-BEING AS EXPERIENCED SATISFACTION OF PRESENT PREFERENCES

There are different ways in which one may try to characterize well-being in terms of preference-satisfaction. The simplest way is by saying that a person's well-being is higher the more his preferences are satisfied and lower the more his preferences are frustrated. However, this account is probably too simple. It gives rise to several problems of which we shall here focus on two.

The first is the problem that when the time comes to fulfil a certain preference the person may have changed his preferences. For example, a lot of people have strong preferences not to become senile when they get old, but when the time comes and they become senile they often don't care. They may be happy amidst their senility. However, given the simple account of well-being in terms of preference-satisfaction it will subtract from a person's well-being that his previously held preference not to become senile is being frustrated. This seems counter-intuitive. The senile person's well-being cannot be affected by the frustration of preferences which he has no longer.

The other problem is that even if one still has a certain preference one may not be aware that it is being satisfied or frustrated. Take a person who for religious reasons is strongly against being given blood. Assume that he is actually given blood during an operation but that the hospital staff manages to convince him that this is not the case. Or assume, conversely, that he has actually not been given blood, but that he is fully convinced that, despite the clear expression of his will, he has been given blood. In the first case the patient's well-being seems to be a unaffected by the fact that one of his present preferences has been frustrated; and in the second case the patient is not made any better off by the fact that his preference not to be given blood, contrary to what he thinks, has actually been satisfied.

To overcome these and other similar problems we suggest characterizing well-being in terms of *experienced satisfaction of present preferences:*

> A person's well-being at a given point in time (t_1) is relative to the degree of agreement between what he at t_1 prefers (wants, aspires after, hopes for) and how he at t_1 sees his situation (past, present and future) — the more agreement the more well-being.

A person's well-being at a given point in time is here characterized in terms of his present preferences and his present perception of his situation. This, however, does not mean that only things happening now will affect a person's well-being at the present time. To see why, consider a phenomenon like regret.

The person's well-being may be affected negatively by his regretting something that he did in the past; and this may be fully accounted for in terms of his present preferences and his present perception of his situation. Thus he *now* wants that he (in the past) did not do something which he *now* perceives himself as doing in the past. A similar account may be given of fear: the person *now* wants to avoid something which he *now* anticipates as coming in the future.

Also it is important to be aware that what a person prefers at a certain time may include life-plans and other preferences which remain constant for long stretches of a person's life. For example, a woman may want a child; if she does not succeed this may affect her well-being aversely most of the time. The suggested account of well-being therefore does not favor short-term satisfactions at the expense of life-plans and other global desires — contrary to what has traditionally often been claimed about subjectivist theories of well-being.

Our account of well-being in terms of experienced satisfaction of present preferences differs from hedonism in one important respect: There is no experienced quality shared by all those things which contribute to a person's well-being. However, in another important respect it is closely related to hedonism: Only things which a person experiences will affect his well-being.

Some philosophers would claim that we have not moved far enough away from hedonism [2], [9]. They would claim that well-being is more independent of our experiences than our account allows for. For example, they would say that a person's well-being is affected adversely by slander, even though the person himself is unaware of the fact that he is being slandered. Similarly, some of these philosophers will probably think that the well-being of the person who becomes senile is aversely affected even though the person himself does not feel less happy.

We shall not discuss these claims. We want only to point out that most of the things we have to say about comparing well-being will be of relevance not only for our own account but for any preference-account of well-being.

COMPARING WELL-BEING: THE SIMPLE CASE

Assume that there is a patient, P, and that there are two kinds of treatment or other kinds of health-care initiative, T(1) and T(2), which he may be offered. One may want to know what must be the case for T(1) to have the best effect in terms of well-being.

Consider the following conditions:

(1)   P's preferences (or at least those of them which are relevant) are known;

(2)   It is known how T(1) and T(2) will affect P's situation;

(3)   T(1) will satisfy at least one preference that is not satisfied by T(2) or will satisfy it in a more effective manner; and apart from this the two treatments have similar effects; and

(4)   neither treatment will affect which preferences P has.

(The expression "satisfy P's preferences" is here used as a shorthand for "affect P's situation so that there is agreement between what he prefers and how he sees his situation".)

When these conditions are fulfilled it may be concluded that T(1) has the best effects in terms of well-being.

A simple example where all four conditions seem to be satisfied is the following: the effect of the drugs omeprazole and cimetidine on gastric and duodenal ulcers has recently been studied [4], [5]. It turned out that omeprazole accelerates healing and pain relief more effectively as compared with cimetidine. No major clinical or biomedical side effects were noted. It is reasonable to assume that most patients with gastric or duodenal ulcers have a strong preference to get rid of their pain and other symptoms as soon as possible and to avoid averse side effects. There seem to be no other relevant preferences. Both treatments will to some extent satisfy the preferences of the typical patient. However, omeprazole will satisfy the relevant preferences in a more effective manner than cimetidine, and apart from this the two treatments have similar effects. Finally, it seems that neither treatment will affect which preferences the patients have. Therefore, it may be concluded that omeprazole as compared with cimetidine (in the relevant doses) has the best effects in terms of well-being.

The example just stated is, of course, not very exciting. All other criteria speak in favor of one treatment compared with the other, and it is then no surprise that this treatment also has the best effects in terms of well-being.

In many cases one may want comparisons of well-being to go beyond what is stated in this simple case. First, one may want to know not only which initiative has the best effects in terms of well-being, but *how much more* well-being comes out of one initiative compared with the other; and one may want to compare the well-being of one patient with the well-being of another. Finally, one may want to compare cases where two health-care initiatives satisfy different preferences.

One influential way of trying to do all these things is by means of the so-called QALY-method.

THE QALY-METHOD

Ill-health may affect a person in two ways: the quality of his life may deteriorate, and he may die. According to the QALY-method these two things are really two dimensions of the same thing. The value of a person's life may therefore be defined as the number of years that he lives multiplied by his level of well-being during these years — hence the expression "Quality Adjusted Life Years".

The value of any health-state may according to the QALY-method be measured by asking people how they rate it compared with two other states the value of which have been defined in advance, "being dead" and "being in full health". Death is assigned the value 0 and a year in full health is assigned the value 1. There are several different ways in which people may be asked to rate the value of a health-state [12], of which we shall here mention only one, the so-called time-trade-off method.

A subject's preference for a certain health-state can be found by asking which number of years in full health he would trade for a number of years in the state of disease. For instance a subject might be indifferent between 5 years in perfect health and 10 years confined to a wheelchair. That gives one year confined to a wheelchair the value 0.5. In principle the value of any health-state can be measured in this way.

Usually the method is not used to measure the well-being of individual persons. Rather, groups of persons are interviewed with the aim of finding average or aggregate values to be used in evaluations of health care programs. This aspect of the method will not concern us here. For critical discussions see [6] [7].

The QALY-method will, if it works, be a very powerful way of measuring well-being. States of well-being will be assigned numbers which can be compared across times and persons. Thus it may be possible to say that the life of a person confined to a wheel-chair has a value of 0.5 compared to his own life before he became handicapped and compared to the life of another person who is in full health.

There are, however, two groups of questions which may be raised concerning the use of the QALY-method as a way of measuring the well-being of individual persons. These are methodological and conceptual questions.

The methodological questions include: How well has the state of ill-health been described? Which other things apart from the relevant health-state should be described, for example his age, his income, family and friends? Has the test-person really identified himself with the relevant situation, or does his rating of the situation only reflect prejudice about what it is like to be ill? Are there other things apart from the expected well-being which may affect the way a person rates certain states of ill-health — for example, that he puts less value on things the further away they are in time?

These questions are, of course, very important to answer for anyone who wants to use the QALY-method. We shall, however, not go into these questions here. Instead we shall concentrate on the conceptual questions.

There are two basic conceptual questions which we would like to consider:

(1)   Does it make sense to ascribe values to how much health-states matter to a person's well-being — for example, the value 1,0 to full health?

(2)   Often a change in health-state will also imply a change in preferences. Does it make sense to rate one's well-being in a possible situation where one's preferences have changed?

We shall discuss these questions in the following two sections.

THE VALUE OF HEALTH

All people seem to prefer health to ill-health, and it may therefore be said that for everyone full health is of value. However, from this it does not follow that full health has the same value to everyone nor that it is possible to assign health a specific value.

Only a little reflection is needed to cast doubt on the assumption that full health will have the same effect on everyone's well-being. Health may be viewed as a resource, something that we can make use of to achieve our goals. If my main aspiration in life is to become a test-pilot, good health is of paramount importance to me. If, on the other hand, I want to become a professional philosopher, it may not matter much that I have to wear glasses and that I suffer from mild hypertension.

The problem for the QALY-method is not just that the philosopher unlike the prospective test-pilot may be unwilling to trade a number of years suffering from myopia and mild hypertension for a significantly smaller number of years in full health. This may just be a sign of ordinary variances in preferences for a certain health state. Rather, the problem is that we have no reason to believe that all persons assign the same value to a year in full health. Therefore, a life year in full health cannot serve as a standard against which (together with death) the values of all other health states are rated.

There is, however, another and more serious obstacle to the idea of comparing how much various health states affect a person's well-being.

Above we pointed out that it made sense to ascribe absolute values to how much it matters to a person to be in certain health-state. However, the idea of ascribing an absolute value to the health of a person must be checked against the above characterization of well-being. For the idea to make sense it must be possible to find some entities of which this value can be seen as a measure. Above we suggested that well-being should be accounted for in terms of preference-satisfaction. Is the satisfaction of a set of preferences something which can be assigned an absolute value?

Before answering this question it may be useful to distinguish between three different kinds of scales for comparing the strength of preferences:

Ordinal-scale

All possible situations are ordered in the following way: for every two situations it is the case either that the person prefers one to the other or that he is indifferent between them.

(Analogy: 1., 2., 3. ... winner in a horse-race.)

Interval-scale

If the person is not indifferent between three situations, it is possible to say how much more he prefers one situation to the two others. For example, it may be possible to say that if one preference has the value 1,0 then the two others have the values 2,0 and 3,0. But since the same may be expressed by saying that one preference has the value 2,0 and the two other the values 3,0 and 4,0, it does not make sense to say that one preference is twice as strong as the other.

(Analogy: temperature-scale without an absolute zero.)

Ratio-scale

It is possible to ascribe absolute values to how strongly a person prefers something. Thus it also makes sense to say that one preference is twice as strong as the other.

(Analogy: linear measure.)

If the satisfaction of a set of preferences is something which can be ascribed an absolute value, then it must be possible to compare our various preferences on a single ratio-scale. However, the whole point of our criticism of hedonism was that there is no one scale on which it is possible to measure and compare well-being between different persons. Let us spell this argument out.

According to hedonism all mental states which contribute to a person's well-being share a certain mental quality, the instances of which differ from each other only by being more or less intense. At least we can make some sense of the idea of arranging all of person's mental states on a ratio-scale according to the degree to which they possess or contribute to the relevant quality of pleasure/pain.

To do this would be like arranging all the various sounds coming out of a stereo-set on a ratio-scale according to the degree to which they contribute to the noise in the room. We can at least understand what is meant by saying that some bit of Rolling Stones is twice as noisy as some bit of Mozart. Similarly, given the truth of hedonism, we can make sense of saying that the pleasure involved in reading a certain philosophical paper is twice as intense as the pleasure involved in eating a certain piece of cake.

However, hedonism is inadequate precisely because there is no one mental quality shared by all those mental states which contribute to a person's well-being. The pleasure of philosophizing (if it is a pleasure) and the pleasure of eating cakes both contribute to a person's well-being; but it does not make sense to say that *one pleasure is twice as intense as the other*. What they have in common is that they are both wanted; and what distinguishes one from the other is that one is preferred to the other. However, since all we have to build on is that one thing is being preferred to the other, we will be able to compare the strength of the relevant preferences only on an ordinal scale.

This allows us to conclude the argument: If it makes sense to ascribe an absolute value to a person's well-being, then it must (at least in principle) be possible to compare how strong a person's various preferences are on a ratio-scale. However, since it is only possible to make such comparisons on an ordinal scale and not on a ratio-scale[1], it does not make sense to ascribe an absolute value to a person's well-being.

The adherent of the QALY-method may reply that he does not want to compare the strength of preferences on a ratio-scale. All he wants is to compare preferences on an interval-scale [11], and since an interval-scale may be constructed on the basis of an ordinal-scale, the presented argument, though valid, will not affect his position — so at least he may claim.

However, we want to argue that an argument similar to the one presented above may be turned against the idea of constructing interval-scales on the basis of ordinal-scales.

All the user of the QALY-method has to build on is the observation that for any two situations a person either prefers one situation to the other or is indifferent between two situations. A test person may for example prefer 10 years in full health to 10 years in a wheel-chair, and he may be indifferent between 10 years in a wheel-chair and 5 years in full health. On the basis of these observations the user of the QALY-method concludes that the values of the three situations, "being dead", "being confined to a wheel-chair" and "being in a state of full health", can be presented on an interval-scale where the difference between the value of the first and the second situation equals the difference between the value of the second and the third situation.

It should be noted that the construction gets under way only because it is *assumed* that there is a given difference between the value of being dead and the value of having full health. This difference serves as a standard of comparison for all the other values which are being compared. However, the question may now be raised whether we can make any sense of there being such a difference. Can we make sense of this just as we can make sense of saying that there is a certain difference between the temperature of freezing water and the temperature of boiling water which may serve as a standard of comparison for temperatures?

The point of the question is not to ask whether there is always the same difference, just as (given certain specifications) there is always the same difference in temperature between freezing and boiling water. Rather, the point is to question the idea of there being *a certain* difference. Have we got any idea of what it means to say that there is a certain difference in the strength of one preference compared with the other which may serve as a standard of comparison for the relative strength of all kinds of preferences?

Here again we want to revert to the criticism of hedonism. The point of this criticism was that there is no quality shared by all the mental states contributing to a person's well-being, apart from the quality of being wanted. However, we do not know what it means to say that there is a certain difference between the strength of one want and the other. All we can make sense of is that one want is stronger than the other. Therefore, it does not make sense to construct an interval-scale for comparing the values of different health-states on the basis of an ordinal-scale, and therefore our conclusion stands, that it does not make sense to ascribe values to *how much* more one health-state matters to a person's well-being than another. All we can meaningfully say is that one matters more than the other.

Some people may object to the above arguments by pointing out that researchers as a matter of fact are able to measure health-related well-being by means of the QALY-method. Therefore, allegedly, it cannot by means of *a priori* reasoning be possible to show that they cannot do so. To try to do so is like arguing *a priori* that there cannot be flying machines.

To disarm this objection we want to remind the reader that the ascription of values to the two states "death" and "full health" is not something that is done by means of the QALY-method. It is itself an *a priori* assumption which has to be made before the QALY-method can be used; and it is in no way being validated through the use of the method. Therefore our arguments cannot be rejected simply by pointing out that they are *a priori*.

This allows us to move on to the other conceptual question raised above.

CHANGING PREFERENCES

In the QALY-method test-persons are asked to rate health-states different from the one they are presently enjoying. They are asked how much it will matter to them if they come to suffer from various kinds of ill-health. Above we discussed whether it makes sense to ascribe a value to how much a certain health-state matters, either by itself or relative to other health-states. Another problem is that people may change their preferences if their health-state changes. The question then arises whether it makes sense to rate one's well-being in a possible situation where one's preferences have changed. This is the question that we shall now go on to discuss.

That preferences can change during disease seems to be confirmed by findings in clinical quality of life research. In one trial approximately 100 women with a diagnosis of breast cancer were interviewed about their preferences for either breast-conserving surgical treatment or mastectomy (surgical removal of the breast). About the time of diagnosis and choice of treatment it turned out that about 2/3 of the women preferred mastectomy or at least did not prefer breast-conserving treatment.

Preliminary finding seems to indicate that, one year after this preference was expressed and the treatment chosen, a significant number of women think differently about their situation, in that more now prefer a breast-conserving treatment[2] The tendency that breast-conserving treatment is eventually regarded as less problematic than mastectomy is confirmed in other findings [3].

There is also evidence that preferences differ between cancer patients and others. Patients with cancer are much more willing to accept a radical chemotherapeutic treatment than people who do not have cancer, including a matched control group, groups of cancer specialists, general practitioners and cancer nurses. Indeed, cancer patients were willing to accept treatment with as little as 1% chance of survival, while the control group would not accept treatment with less than 50% chance of survival [11].

Some of these findings may, of course, reflect differences in knowledge about what it is like to be ill; but there is no reason to think that they all do. In some cases it seems that *illness makes people undergo radical changes concerning what matters to them*. For example, very ill people often seem to value some of the basic things in life much more highly than people who are not ill.

To see how this bears on the QALY-method consider the following example: Two health-scenarios are presented to a test-person. In each case his preferences are to some extent being satisfied. However, his preferences will not be the same in the two cases. Is it possible to say which scenario is preferable?

It may be thought that the person faced with the choice could simply rate the options on the basis of the preferences he presently has. However, this would be to evade the problem. In the two cases the person will have sets of preferences which differ from each other and from the ones he now has. What determines the person's well-being in the two cases are his new preferences and the extent to which they are satisfied, not the preferences he has at the moment of choice.

It may be a fact that people faced with the relevant kind of choice unhesitatingly choose on the basis of their present preferences. However, this is a mere psychological matter and has nothing to do with deciding the well-being in the two alternatives.

As another way of getting round the problem, it may be suggested that the alternatives should be compared directly: how much well-being results in each case measured by the preferences that the person will have in each case? However, this strategy presupposes that it is possible to ascribe values to the amount of well-being that comes out of each scenario, and this possibility has been excluded by our arguments in the previous section.

Finally, it may be thought that the well-being in the two options can be rated from the point of view of one of the cases. In one alternative the person develops a set of preferences, and on the basis of this both alternatives could be compared. But of course this will not do. On the basis of the preferences of the first scenario one might express a preference for the second scenario; but this will say nothing about the level of well-being in the second scenario because this depends on the preferences the person has in that second scenario.

We may therefore conclude that when a person undergoes such changes in preferences as described above there is no longer one perspective from which to rank the different health-states involved — unlike in the simple case presented above. In the simple case the preferences remain the same and therefore it is possible to compare the different situations according to the degree to which they satisfy the person's preferences. If furthermore, as in the simple case, one situation satisfies all the preferences that the other does and some more, then that situation may be said to have the best effects in terms of well-being.

Comparisons of well-being presuppose a set of preferences which serve as a standard of comparison. If two situations involve different preferences there is no standard of comparison on the basis of which these two situations can be compared. Therefore, it does not make sense to rate one's well-being in a possible situation where one's preferences have changed. Let us call this conclusion the *incommensurability thesis*.

However, even if it strictly speaking does not make sense to rate the well-being of a person in two such situations, it may be possible to make other comparisons which will at least serve as a reasonable substitute for a comparison of well-being.

OTHER COMPARISONS

A person is healthy and much prefers to stay in this condition. One day he becomes ill. He cannot be cured and therefore has to live with certain chronic states of ill-health. He manages to some extent to adjust to the situation by a change of aspirations. However, he would still much prefer to be in full health.

Since the person has changed some of his preferences, it does not make sense to compare his well-being in the two situations, but it is still possible to make another interesting comparison. Both from the point of view of his previous and his present situation being in full health is much to be preferred to the relevant kind of ill-health. And therefore full health may be said to be *preferable from all relevant points of view*.

To find out that something is preferable from all relevant points of view may in many cases be all that people are interested in when they are trying to compare health-related quality of life. Therefore the incommensurability thesis may not be a major obstacle to quality-of-life research.

There are, however, cases where the incommensurability thesis applies, where health is not preferable from all relevant points of view and where it is tempting to think that it should be possible to make some kind of comparison of well-being. Take a person who becomes senile. Before, he like most other healthy persons had a strong aversion against becoming senile; but later in a state of fully developed senile-dementia he has no preference for regaining his previous mental capabilities, and he seems by all normal criteria to be content with his situation.

Is there any other way in which one may compare the well-being of a person in two situations like these? One way could be to expand the concept of well-being.

According some preference-accounts of well-being, the quality of a person's life depends on how much agreement there is between his preferences and his situation (or his situation as he sees it) — the more agreement the more well-being. Many people will agree with this as far as it goes, but say that there is more to well-being than just agreement. Thus Lennart Nordenfelt ([8], p. 52) argues that an extra dimension of well-being (happiness) should be added to the dimension of agreement (equilibrium):

> Consider a boy who has lived all his life in simple and unpretentious circumstances in the Highlands of Scotland. He has been entirely content with his lot; he gets along well with his family, he appreciates the wild landscape and he enjoys the sometimes hard struggle for life up there. Thus, for a long time he has been completely happy in our want-equilibrium sense.
>
> One day he and his family are visited by a tourist who happens to be a famous musician. The tourist is attracted by the place and settles there permanently. Partly to earn his living he starts teaching the young boy to play the violin. He then discovers that the boy has a remarkable talent for music and he very soon develops a proficiency in playing. This completely changes the boy's life. A whole new world has been opened to him and he acquires a set of completely new goals in life. To put things in my technical terms, he has acquired a multitude of new wants, which he didn't have before, and he is in the process of satisfying them.
>
> The boy was, as we said, completely happy with life before the musician arrived. But how should we then express the positive change that has now happened to him? Let us here presuppose that it is a positive change. One can, of course, imagine the case where the appearance of the musician disturbs the idyll in the family. The boy's parents may become envious or even jealous; they may want to get rid of musician and thereby ruin the boy's life.
>
> But the case that we shall consider is instead the following: The boy has all his previous wants satisfied; in addition he now has a number of further wants; some of these, or even all of them, are being satisfied. In quasi-numerical terms, the boy may now have 20 out of 20 wants satisfied, instead of 10 out of 10 wants satisfied.
>
> In order to characterize this case I shall now introduce a further dimension of happiness, which I shall call the *dimension of richness*. The Scottish boy is now

happier than before in the dimension of richness, but strictly speaking equally happy as before along the equilibrium dimension.

Richness does not solely concern an increase in the number of wants. It can also entail the case where some modest wants are replaced by a set of more *ambitious* wants.

If Nordenfelt is right in his claim that richness is a separate dimension of well-being/happiness, it is clearly possible to compare the well-being of a person when he is healthy and when he has become senile. Compared in the dimension of richness, suffering from senile dementia is surely a less happy situation than possessing one's full mental powers.

Nordenfelt's suggestion will also have important consequences in other kinds of cases.

Often it is not possible to cure patients. They will have to live with certain chronic states of ill-health. For example, after an acute myocardial infarction a patient may due to myocardial failure be able to do much less than he could before. The only way to improve the well-being of such a patient is to make him adjust his preferences so that they are in accordance with his new situation. By becoming less ambitious he may be able to restore the agreement between his preferences and his ambitions.

However, since some relatively ambitious wants are here being replaced by a set of more modest preferences it seems that according to the dimension of richness, the situation of the patient must deteriorate. And therefore, if Nordenfelt is right, the patient loses some quality of life even if full preference-satisfaction is achieved.

Nordenfelt may reply that the dimension of *richness* is not just a matter of having more preferences or having more ambitious preferences. For a life to count as more rich the relevant preferences should also be fulfilled. Since the heart patients are not able to fulfil their previous more ambitious preferences their lives do not become less rich when they become less ambitious. However, according to Nordenfelt, it still remains the case that in a sense these patients are less happy than they used to be because their lives are less rich. This we think is counter-intuitive.

According to our view, richness is no independent dimension of well-being; and the example which pulls the weight in Nordenfelt's presentation can be accounted for without revising the account of well-being as experienced preferences-satisfaction presented above.

The Scottish boy seems to be the kind of person who takes pleasure in doing and achieving things. Before he met the musician he was not aware that there are other ways of doing this than by struggling along in the rural life of the Scottish Highlands. Through the musician, however, he becomes aware of new and more challenging ways of achieving things. The increase in well-being is not caused by the fact that the boy now satisfies a desire which he did not have before. Rather he becomes aware of new and better ways of satisfying preferences that he has all along had, e.g., the desire to do admirable things, and this improves his life.

A similar manoeuvre cannot be used to account for the example of the person suffering from senile dementia. The senile person no longer has aspirations in the light of which one may say that he would be better off with his previous mental capabilities. To say that the senile person has a less good life because his life has become less rich is, according to our view, an unwarranted *ad hoc* manoeuvre. According to our view, one simply has to accept that in some cases the incommensurability thesis applies and that some health states are not preferable from *all* relevant points of view.

Luckily, most cases are not like the case with the person suffering from senile dementia. In most cases, to be more healthy is preferable from all relevant points of view. Furthermore, many cases will be sufficiently like the simple case presented above for comparisons of well-being to make sense. Therefore, even if we are right in all our skeptical conclusions, there is still room for many interesting comparisons of health-related quality of life[3].

NOTES

[1] It should be noted that we are not saying that it is never possible to compare the satisfaction of two preferences on a ratio-scale. It may for example be possible to compare two experiences of pain on a ratio-scale. What we are saying is that there are kinds of preferences (the preference for philosophy and the preference for a piece of cake, e.g.) which cannot be compared in this way.

[2] Personal communication from Anders Bonde, Department of Oncology, Odense Hospital, Odense, Denmark.

[3] We want to thank Roger Crisp, Nils Holtug, Sven Erik Nordenbo and Jesper Ryberg for helpful comments on an earlier version of this paper; thanks are also due to the Danish Research Council for the Humanities for its financial support.

BIBLIOGRAPHY

[1]   Bentham, J.: 1789, *An Introduction to the Principles of Morals and Legislation*, quoted from John Stuart Mill, *Utilitarianism*, Collins/Fontana, 1962.

[2]   Griffin, J.: 1986, *Well-Being*, Clarendon Press, Oxford, ch. 1.

[3]   Kiebert, G.M. *et al*: 1991, 'The Impact of Breast-Conserving Treatment and Mastectomy on the Quality of Life of Early-Stage Breast Cancer Patients: A Review', *Journal of Clinical Oncology, 9*, 1059-1070.

[4]   Lauritsen, K. *et al*: 1985, 'Effect of Omeprazole and Cimetidine on Duodenal Ulcer', *New England Journal of Medicine, 312*, 958-961.

[5]   Lauritsen, K. *et al*: 1988, 'Effect of Omeprazole and Cimetidine on Prepyloric Gastric Ulcer: Double Blind Comparative Trial', *Gut, 29*, 249-253.

[6]   Loomes, G. and McKenzie, L.: 1989, 'The Use of QALYs in Health Care Decision Making', *Social Science and Medicine, 28*, 299-308.

[7]   Nord, E.: 1992, 'Methods for Quality Adjustment of Life Years', *Social Science and Medicine, 34*, 559-569.

[8]   Nordenfelt, L.: 1994, 'Towards a Theory of Happiness: A Subjectivist Notion of Quality of Life', in this volume, pp. 35-57.

[9]   Nozick, R.: 1974, *Anarchy, State, and Utopia*, Basil Blackwell, Oxford, ch. 2.

[10]   Sidgwick, H.: 1907, *The Methods of Ethics*, 7th ed., reprinted by Hackett Publishing Company, Indianapolis/Cambridge, 1981, book II, chs. 1-3.

[11]   Slevin, M.L. *et al*: 1990, 'Attitudes to Chemotherapy. Comparing Views of Patients with those of Doctors, Nurses, and General Public', *British Medical Journal, 300*, 1458-1460.

[12]   Torrance, G.W.: 1986, 'Measurement of Health State Utilities for Economic Appraisal. A Review', *Journal of Health Economics, 5*, 1-30.

A.W. MUSSCHENGA

QUALITY OF LIFE AND HANDICAPPED PEOPLE

SUMMARY

In the first four sections of this paper, an analysis is made of the meaning and the of the term 'quality of life' in medicine and health care. It is shown that the term refers to different concepts such as normal functioning, satisfaction with life, and level of human development. The term was at first no more than a vehicle for critique on a medicine that was fixated on prolongation of life and containment of disease processes. Later it was used to refer to more positive criteria for evaluating the medical effectiveness and the economic cost-effectiveness of treatments and other health care provisions.

In the last section the relevance of quality-of-life-considerations for the care for the handicapped is discussed.

1. CULTURAL BACKGROUNDS OF THE USE OF THE TERM

The origins of the term 'quality of life' lie outside the domains of medicine and health care. As far as I know, the first to use the term were Samuel H. Ordway [18] and H. Fairfield Osborn [19]. In the beginning of the 1950s they wrote books in which they indicated the dangers of exhausting non-renewable resources and questioned the belief in technological and economic progress. Ordway warned that the dominant ideas of the good life and values had to be revised. Failing that, the end of industrial expansion could well become the end of our civilization. The term 'quality of life' also plays a role in the writings of the well-known economist, John Kenneth Galbraith [8] [9].

So the early use of the term 'quality of life' is connected with a critique of a certain conception of the good life in which material values figure prominently. The critique stresses the devastating long-terms effects (exhaustion of resources) and side-effects (pollution of the environment) of economic growth on the environment and on the future conditions for a good life. In the 1970s this critique was widely disseminated, especially by *Limits to Growth*, the Report for the Club of Rome [17]. Not only for that reason did Ordway criticize the dominance of material values, but also because of the underlying reductionist image of humanity. The exhaustion of

181

L. Nordenfelt (ed.), Concepts and Measurement of Quality of Life in Health Care, 181–198.
© 1994 *Kluwer Academic Publishers. Printed in the Netherlands.*

the natural resources and the degradation of the environment go together with the degradation of humanity as little more than consuming being. As Herbert Marcuse formulated in 1964, man kind has become one-dimensional [13]. Although authors like Marcuse and Erich Fromm [6] [7] do not use the term 'quality of life', it is clear that their critique of a materialistic culture closely resembles that of the environmentalists.

The term 'quality of life', orginally expressed a concern for the *quality of the conditions for living,* which are threatened by the exhaustion of natural resources and the deterioration of the environment, processes which endanger the very survival of the human species. In the critique on the narrow view of the good life which is driving the hunt for material welfare, the term also expresses a concern for the *quality of being human.*

As long as the term was simply a vehicle for critique, there was no need for a more precise definition. That changed when in the 1970s the improvement of the quality of life became an independent *political* goal, besides the maximization of welfare. In order to know how the quality of life can be improved, one has to discover which factors exert a positive or a negative influence. That can only be done by *measuring* the quality of life under different conditions. If there is no clear definition of the term, even on operational one, it is impossible to design instruments for a valid measurement of quality of life. Discussions about all kinds of conceptual and methodological questions pertaining to quality of life continue to the present. To trace its course, one only has to study the articles in the *Journal of Social Indicators.*

Medicine and health care are just two important factors — surely not the only ones — which may influence the quality of life — in the sense of *conditions for living.* One can imagine that social policy makers were interested in the role medicine could play in improving the quality of life. However, that is not the reason why the term became popular in medicine. Its popularity is due to its usefulness for internal discussions about the relevance of medical treatments, and of medicine in general. Already in 1966 in an editorial in the *Annals of Internal Medicine* (under the title "Medicine and the Quality of Life"), J.R. Elkinton discussed the need to allocate the health care budget in such a way that it delivers a maximum contribution to the "health and quality of life of all members of society". In connection to that, he points at the ambivalent effects of very invasive treatments eg. kidney dialysis and kidney transplantation. In the socio-political domain the term 'quality of life' is a vehicle for critique of economic growth and material progress. Analogously, in medicine it becomes the vehicle for critique of expanding medical power and technical medical progress. Nowadays 'quality of life' has become a standard term in medical circles, while its use outside the domain of medicine is infrequent.

2. THE MEANING OF THE TERM: A FIRST EXPLORATION

The impact of the term 'quality of life' cannot be explained by its descriptive quality only, because it is vague. It is rather the other way around: The vagueness and open boundaries of the concept did contribute to its usefulness. However, that cannot be the whole explanation. Judged by these criteria, a much older notion — "well-being" — is as useful as "quality of life". What are the advantages of the term 'quality of life' beyond that of "well-being"? The most simple explanation is that the opposite of "well-being" in the socio-political domain, "(material) welfare" is not applicable within the domain of medicine. The pair of concepts "quality of life" and "quantity of life" was found to be appropriate for both characterizing and criticizing the dominant aims and goals of medicine. The core of the critique of medicine was that it measured its success and progress in purely quantitative terms. Dominant medicine was, according to its critics, fixated on increasing quantity of life. The term 'quality of life' was used to refer to aims and values that were neglected in modern medicine. If we can clarify the meaning of the concept of quantity of life, we might expect to be a step nearer to the identification of the meaning of quality of life.

Quality of life is something that is different from and in contrast to quantity of life. Until now we did not know much more than that. The meaning of the term 'quality of life' depends on the nature of its relation with the term 'quantity of life'.

The core meaning of quantity of life is something like length of physical life length of physical survival, or average life expectancy. As the opposite of the term 'quantity of life', quality of life can either be a descriptive or an evaluative term. As a descriptive term, quality of life refers to the defining characteristics of the biological species *homo sapiens*. The qualities of a knife, a table, a human being, etc. are those properties that constitute their identity as a knife, table, or human being or manager. The qualities of an entity mark the boundary between that entity or social role and other entities or social roles. As an evaluative term, quality refers to excellence or flourishing. A blunt knife has no quality, but it is still a knife. An X can only be called a qualitatively good X — in the normative sense of quality, if it has the qualities of an X — in the descriptive sense of quality. If we speak about the quality of someone's life, we thereby mean the quality of his life as a human being. However, there is no tight watershed between a normative and a descriptive use of the term "quality". The statement that the life of a being that is born out of a woman, does not show the distinctively human qualities, is also secondarily a negative value judgment, for the simple reason that such beings are expected to have those qualities.

The term 'quality of life' is most frequently used in the normative sense. However, as I will show, one meaning of the term in medicine is not derived from the normative, but from the descriptive concept of quality of life. For the moment I will presume that there is a consensus about the defining characteristics of the human species. Is there also a consensus about human excellence and human flourishing?

The concept of human excellence is connected with classical Greek philosophy especially the ethics of Aristotle (384-347 B.C). The Greek term for human excellence is *arete* (virtue). Aristotle in the *Nicomachean Ethics* distinguishes between different kinds of excellences: bodily excellences, intellectual excellences, and excellences of character [1]. A life in which all human excellences are given full expression, is called "eudaimon" by Aristotle. Eudaimonia or happiness is the highest and most complete good, the highest quality a human life can acquire. In my opinion the current meaning of the term 'quality of life' is closely related to that of happiness.

3. EUDAIMONIA AND QUALITY OF LIFE

The problems one meets in identifying the meaning of quality of life can be illuminated by tracing the development of the meaning of the concept of happiness. There are several differences between the Aristotelean concept of eudaemonia and any modern concept of happiness. First eudaimonia is a moral concept. Aristotle does not distinguish between happiness and moral character. The happy man is also the good man. Although virtuous action does not guarantee happiness, since there is always an element of luck or fortune, it is a necessary condition for happiness. In modern ethics one does distinguish between happiness and moral goodness. For us it is an observational fact that crooks can become happy as a result of their evil actions. Second "eudaimonia", unlike modern "happiness", is not only a moral, but also an objective concept. The paradigm for eudaimonia in Aristotle is health. Third Aristotle does not as clearly distinguish between happiness and conditions of happiness as we would. Wealth and health are for Aristotle not just conditions of happiness, they are elements of it.

In modern ethics — especially in utilitarian theory — the term 'happiness' does not refer to an objective state of affairs. In classical utilitarianism it does refer to the subjective, inner *experience* of the satisfaction of desires; to positive mental states. In more contemporary utilitarian theories happiness is not a subjective experience, but a subjective, first-person *evaluation*. A well-known modern definition of happiness is "satisfaction with life". (I will return to this definition later on.) Satisfaction with life is a function of the distance between someone's ambitions and aspirations on the one hand, and his capacities and opportunities on the other. The smaller the gap, the greater the satisfaction with life.

Of course there is a relation between happiness as satisfaction with one's life and certain objective states of affairs. Health and wealth, for example, are generally regarded as conditions for happiness. However, these conditions are neither sufficient nor necessary for happiness. In most cases health and wealth do contribute to happiness. That is why they can be seen as indicators of happiness. It is perhaps more difficult for ill and poor people to become happy, than it is for healthy and rich people. Notwithstanding that, many rich and healthy people are unhappy, while many poor and ill people are happy. Health and wealth do have instrumental value for happiness, but they can be absent. At the same time they also have intrinsic value. They are desirable for their own sake. Happiness can refer to diverse configurations of values. Only in some of them do the values of health and wealth have a prominent place. A physically handicapped man can be happy, if he manages to accommodate his aspirations to the limitations of his handicap. Does that imply that this man is indifferent to his handicap? If he should say so, he probably would not be understood, even by other handicapped people. But if he is happy, why should he long for complete health? Can a happy handicapped become more happy if he is cured or regains lost functions? If that person declares that he is more happy after being cured, should one not conclude that he was not really happy when handicapped?

What did we learn from this intermezzo about the concept of happiness? As we have seen, in ordinary language one of the basic meanings of quality of life is human excellence. Eudaimonia in Greek ethics is the fullest expression of all human excellences. In the context of medicine (which I explore in more detail in the next section) this meaning of human excellence is still present, but it is not dominant. The dominant meaning comes very near to the modern utilitarian concept of happiness as satisfaction with life. However, the actual use of the term 'quality of life' is much more ambiguous than that of the utilitarian term "happiness". Sometimes the term 'quality of life' refers to satisfaction with life, but sometimes to the conditions that contribute to happiness.

It should be clear by now that there is no single meaning of *the* concept of quality of life. We distinguished between descriptive and evaluative uses and, within the class of evaluative concepts, between objective and subjective one's. The meaning of the term depends on the context in which it is used.

4. QUALITY OF LIFE IN MEDICINE

The use of the term 'quality of life' in medicine and health care is a *negative* one: to question the role of biological parameters like survival or reduction of the size of tumors as the sole criteria for the effectiveness of medical treatments. The term 'quality of life' then refers to other factors which have to be taken into account in making medical decisions. Which factors, will depend on the nature of the decision that has to be made. That is why it is understandable that there are different kinds of quality-of-life-considerations, related to different concepts of quality of life. The preference for meaning depends on the kind of *decision* one has to make.

But whence the need of quality-of-life-considerations? Until a couple of decades ago, infectious diseases were the main cause of death. The discovery of penicillin and streptomycin, however, made it possible to treat diseases as pneumonia and tuberculosis effectively. In the great majority of cases patients regained complete health following treatment with these medicines. There seldom was uncertainty, either about the probability of a treatment's success, or about completeness of recovery. After gaining control over infectious diseases, not only the relative, but also the absolute incidence of other diseases, (lethal ones like cancer and cardiovascular diseases and chronic ones like rheumatoid arthritis) increased.

In the course of time there were some successes in developing treatments for these diseases, especially cancer and cardiovascular diseases. However, in many cases the effects of such treatments are limited uncertain and temporary. Limited, because the patient's health is not completely restored and he remains hadicapped. Uncertain, because the success and the degree of success of a treatment cannot be predicted. Temporary, because the diseases can recur again, perhaps in other parts of the body, as with cancer or arthritis. Moreover, when chemotherapy for cancer is initiated there is the heavy burden of side-effects of the treatment: e.g., nausea, vomiting, loss of hair, etc. Because of the uncertain, limited and temporary character of the success of many treatments, length of survival is no longer a sufficient measure of treatment success. The term 'quality of life' is the common denominator for a diverse set of other criteria of clinical success. The success or effectiveness of treatments is relevant (as I hope to show later on), not only for decisions about the treatment of individual patients, but also for policy questions, such as the choice of standard treatments and the allocation of scarce resources.

In order to uncover the different kinds of quality-of-life-considerations and the different concepts of quality of life, it is useful to examine their role in the different contexts of medical decisionmaking. In medicine there is not only an ongoing debate on the meaning of the term 'quality of life', but also on the moral permissibility of the use of quality-of-life criteria for prolonging life. According to the moral critics, the use of quality-of-life considerations conflicts with one or more of the principles of respect for autonomy, respect for life, and equality or equal value of all human life. Let us distinguish four contexts:

1. comparative evaluation of the effectiveness of alternative treatments (e.g., kidney dialysis and kidney transplantation);
2. choice between different treatments on the level of the individual patient;
3. decisions about the initiatong or foregoing life-sustaining treatments; and
4. decisions about the allocation of scarce health care resources.

In this article I will confine myself to the first three contexts.

4.1. The choice of a (standard)treatment

In this section, I will take the first two contexts together. The concern for the ambivalent character of, for example, chemotherapy, dates from before the time the term "quality of life" became popular. Already in 1948, D.A. Karnofsky and J.H. Burchenal called attention to problematic features of a nitrogen mustard treatment [10]. They were the forerunners of a tradition of medical evaluation research. In the beginning of the 1970s the term 'quality of life' became more popular. The term then covered a rather loose set of indicators, relevant for measuring *normal functioning and independence.*

Normal functioning and independence were regarded as relevant for the
well-being of patients. There was hardly any interest in conceptual questions
and in the validity and reliability of measurements. The emphasis on normal
functioning fits into an objective concept of well-being.

Since the beginning of the 1980s there was an increase in the participation of social scientists in evaluation research. The social scientists had a
keen interest in conceptual and methodological questions. They brought with
them their own concepts and methodology which stemmed from the tradition
of socio-political quality-of-life research. The dominant concept of quality
of life in that tradition was the one developed by A. Campbell *et al.* in *The
Quality of American Life* [4]. Quality of life refers to the *subjective*
evaluation and experience of life. This evaluation is a function of the
distance between a person's aspirations and their realization. The final
measure is "*satisfaction with life as a whole*" or "*overall-satisfaction*". A
person does not need to be satisfied about each aspect of his life to be
satisfied with his life as a whole.

The aim of this quality-of-life research was to collect data which could
be relevant for the improvement of the effectiveness of treatments, the
comparison of the effectiveness of alternative treatments and decisions about
the treatment of individual patients. The debate concerning a subjective and
an objective concept of quality of life is interesting for several reasons. Before discussing this debate some remarks on the meaning of the term
'subjective' are necessary.

The term 'subjective' sometimes refers to intrapsychic states of affairs,
like feelings, moods, emotions, etc. Although there is a strong correlation
between certain behavioral signals and intrapsychic states of affairs, a person
himself is the most reliable source of information about his feelings etc. The
term 'subjective' in 'subjective evaluation of the satisfaction with life' does
not especially refer to subjective factors, but to a point of view or perspective. A judgment about the satisfaction with life is necessarily a first-person
judgment, a judgment from the point of view — the values, aspirations — of
a person himself. In my view there is no need to choose between a subjective or an objective concept of quality of life, for these concept are relevant
and appropriate, depending on the kind of questions that have to be
answered.

Regarding the subjective concept it is obvious that in many situations the
decision between alternative treatments or between treatment and non-
treatment strongly depends on idiosyncratic factors. Not only are there differences between people in their ability to cope with handicaps, there are

also differences in the impact of handicaps on a person's life-plan. A professor who is confined to a wheelchair following car accident, will still be able to carry out functions which are important for him. Contrary to this professor, a sportsman in the same situation will virtually have to restructure his entire life-plan. Decisions about treatments which are burdensome, while the success is uncertain, limited and temporary, can only be taken from the perspective of the patient himself. Furthermore, it has become clear from several researches that the main determinant of (subjective) quality of life after an invasive treatment is not, for example, the severity of the handicap or the level of normal functioning, but coping ability.[1] The knowledge that the adjustment to and acceptance of less-than-optimal outcomes of treatments is influenced by personality traits, can be useful in deciding whether or not to provide a treatment to a particular patient and in determining the need for psychosocial care. It is not wise to give a pessimistic patient with few coping abilities and a low tolerance for uncertainty an experimental treatment like a liver transplant. He will probably not be able to live well given the risk of rejection of the new liver. The patient's own point of view is also relevant, e.g., in deciding about a treatment that probably will result in some years of survival, but with severe handicaps and burdensome side-effects. He himself then has to make the trade-off between the value of length of survival and that of normal functioning. A good example is the choice that patients with cancer of the larynx sometimes have to make between laryngectomy and radiotherapy. Laryngectomy offers the best chances for survival, but destroys the capacity for normal speech. With radiotherapy that capacity is not affected, but the chances for survival are lower. A study by B.J. McNeill, et al. [15] has shown that some people for that reason prefer radiotherapy, while others have a lesser problem with adjusting to alternative forms of speech, e.g oesophageal speech.[2]

Is there still a place for an objective concept of quality of life? As was already known from the socio-political quality-of-life research, there is only a loose connection between objective conditions and satisfaction with life. As has been shown in some studies, many of the people who were (not) satisfied with their life before a treatment, were also (not) satisfied with their life after it. Even invasive treatments need not change patient's subjective quality of life, neither for the better nor the worse. A patient's judgement of his subjective quality of life after a treatment tells us more about his personality and character than about the quality of the treatment. But the conclusion drawn cannot be that therefore objective conditions — e.g., in medicine e.g. length of survival and level of normal functioning — are irrelevant. In almost every case patients — whether or not they have adequate coping abilities — will prefer a treatment which results in a longer life with less handicaps over one that results in a shorter life with handicaps. The aims of medicine — restoration of health, maintenance or restoration of normal functioning, and alleviation of pain and suffering — do not include the improvement of satisfaction with life as such. In many situations it is sufficient to know what will be the effects of a treatment on length of life and normal functioning. Another limit to the relevance of subjective quality-of-life judgements for medical decisions becomes clear when one realizes that such judgements are not always rational and well-informed. As is pointed out by A. Cribb [5] in his response to K.C. Calman [3], it is possible that a patient's subjective quality of life decreases, while as a result of a medical treatment his physical and mental capacities improve. Such a discrepancy may be due to a temporary depression. In such cases doctors tend not to discontinue treatment because the subjective quality of life decreases. My conclusion is that neither the subjective, nor the objective approach in quality-of-life-research should be neglected.

Doctors are expected to assist patients and, above all, first, to do no harm (*primum non nocere*). If quality-of-life considerations can be useful to obtain a clearer picture of the benefits and burdens of treatments, there can hardly be any objection against their use in treatment decisions. Some critics, (e.g., the Dutch health lawyer Leenen), point to the fact that a judgement about quality of life is a first-person judgement [12]. That is why, in their view, doctors should not take quality-of-life considerations into account in treatment decisions. That objection is misconceived. Doctors are well equipped to determine the effects of treatments on normal functioning and independence, even in cases in which patients are totally or temporarily incompetent. Quality of life as satisfaction with life is, as I noted, a first-person judgement. Each person is the only source of information about his satisfaction with his life. He will probably produce that information only if he is confronted with the right questions. If the methods a doctor uses in obtaining information about the possible impact of a treatment on someone's satisfaction with life are reliable, there can be no objection to using such information in making his decisions. More positively formulated: Decisions based on subjective quality-of-life considerations will probably do justice to a patient's own perspective and his autonomy. However, critics like Leenen have a point. Not all patients are able to orally articulate verbally a judgment about the quality of their life. It is unjustified for a doctor to act upon his own ideas about such patients' satisfaction with their lives.

4.2. Non-treatment decisions

The third context concerning decisions about foregoing life-sustaining treatments. In this context the term 'quality of life' is used in very different senses. In most cases the term refers to a normative — subjective or objective — concept; in some cases, however, it suggests a descriptive concept of quality of life.

Although there is a close analogy between the choice of a treatment from a range of options and a choice between treatment and non-treatment, an important change in perspective takes place. Normally, in all medical decisions there is a tacit presumption in favor of the prolonging of life. The question that has to be answered is which treatment is best in a particular case. However, sometimes the condition of the patient is so poor, that even the best available treatment is benefucuak. In that situation a choice has to be made — not between alternative treatments, but between a treatment with an uncertain or temporary, outcome and non-treatment with the certainty of death.

The moral difference between normal treatment decisions and decisions about or forgoing treatments or active euthanasia is, of course, that in the latter the question of life and death is paramount. Pro Lifers have serious objections to the use of quality-of-life-considerations in decisions about preferenting treatment or non-treatment. Decisions to provide life-sustaining treatment should be based only on a consideration of the probability that it will indeed result in a reasonable prolongation of life. The use of quality-of-life-considerations is in conflict not only with the duty to respect life, but also with the principle of the equal value of human life. Is their critique justified?

In my view, one can hardly deny that normative — subjective or objective — quality-of-life considerations are relevant for a choice between treatment or non-treatment. Treatment decisions, however, should not only be based on the likelihood of their life-prolonging effects. Some patients prefer death over life with a severe handicap, while others in the same situation want to continue to live. Medical criteria are not decisive for such decisions. However, if the principle of respect for life is interpreted as an absolute duty, there is of course no room for the use of any kind of quality-of-life considerations — neither subjective, nor objective — in decisions about life and death. A trade-off between length of life and quality of life or well-being is, in this case, never admissible. Respect for life dictates choice for the alternative with the best chances for survival. In the view of the adherents of a "respect-for-life ethics" the use of quality-of-life criteria is an offense against the sanctity of life, its intrinsic value, the equal value of all human life. This view is due to a misunderstanding of the use of quality-of-life-criteria. They are not used to rank the value and worth of individual human lives, as some critics seem to think, but to assist in weighing the benefits and burdens of treatments.[3]

In discussions about criteria for non-treatment-decisions, still another kind of quality-of-life consideration is found, different from the normative, subjective, or objective ones. A good example is the case of treatment of patients in a persistent vegetative state, (PVS). Although their sleep-and-wake rhythm is normal, they are permanently unconscious. That means that *they* do not have any experiences, negative or positive. They can neither be benefitted nor be harmed by any medical action; neither by prolonging their lives, nor by letting them die. Moral principles such as non-maleficence, beneficence, and respect for autonomy are not applicable. In a narrow sense continuation of treatment (feeding, respiration) is not futile, because it does prolong life. A decision to stop treatment can only be based on a judgement about the value of permanently unconscious life.

When does life have value? That is a very difficult question. We frequently pass judgements about the value of the lives of our fellow human beings: based on their contribution to society — their social value — and the moral quality of what they are doing with their lives. Such value judgments are irrelevant for medical decision making. In my view, the life of a human being has value if he or she has or can have experiences which are judged worthwhile. A person whose neocortex does not function, cannot have any experiences at all. His life has no centre, no subject. Therefore it has *no value*. If the neocortex does not function, the biological precondition for developing distinctively human qualities is absent.

Let me summarize. My analysis has shown that there are different kinds of quality-of-life considerations, based on different concepts of quality of life. The one's I have discussed are:

1.   quality of life as (the degree of) normal functioning (as a member of the biological species homo sapiens) or objective well-being;
2.   quality of life as (the degree of) satisfaction with life or subjective well-being;

The two kinds of quality-of-life considerations are used in weighing the benefits and burdens of treatments. Their use is not problematic from a moral point of view, unless one sees prolongation of life as an absolute duty.

5. QUALITY OF LIFE AND THE HANDICAPPED

What is the relevance of quality-of-life considerations in the context of the care of the handicapped? The discussion about the meaning of the term 'quality of life', about the methodology of its measurement and the moral permissibility of the use of quality-of-life considerations, has concentrated on decisions about medical, primarily curative, life-prolonging treatments. These medical questions are not primarily the focus of interest in research on the quality of life of the handicapped. Its aim is to acquire more insight into factors that can contribute to the improvement of, e.g., the care of the handicapped, their independence, their education. I will just say a few words about the relevance of my conceptual clarifications to that kind of quality-of-life-research. After that, I will concentrate on the question whether quality-of-life considerations show a negative bias towards the handicapped, when used in the context of decisions about life and death.

5.1. Care of the handicapped

The handicapped are not a distinct subspecies of the species *Homo Sapiens*. "The handicapped" is a label for people who have one or more physical and/or mental disorder(s), that differ as to severity, cause, etc. It goes without saying that there is an enormous difference between a person with Down's syndrome, who is mentally retarded and has a reasonable life-expectancy, and a person with Duchenne's muscle dystrophia, who is not mentally retarded and probably will die before he is twenty and whose condition deteriorates with the passing of the years. Perhaps the only common problem of the handicapped is that even in their adult life, they will never become as independent of others as "normal" people. Not only do categories of handicapped persons differ from each other in many respects; all the handicapped, also those belonging to the same category, are in the first place individual personalities with distinctive characters.

Very much research is and has been done which contributes to a greater understanding of the development of the handicapped, their needs, capacities and potentials and so on. For example, no one will dare to say nowadays that even the severely mentally handicapped are uneducable. Much of the research on the handicapped can rightly be labelled as quality-of-life research.

I think it makes sense to distinguish also within research on the handicapped between objective and subjective quality-of-life research. Much research is done which can lead to a greater adaptation of the environment of the handicapped to their special situation and needs. I would characterize that kind of research as objective quality-of-life research. All kinds of new technologies improve handicapped people's possibilities of hearing, seeing, moving, learning, communicating and having jobs. New rules and policies increase the accessibility of public places and buildings such as offices, shops etc. The guiding idea is that the handicapped should be able to lead a life as normal and independent as possible. Leading a normal life not only strengthens a handicapped person's self-confidence, it also counteracts stigmatization by the non-handicapped. That is the truth behind the normalization principle which was so popular in the seventies.

Many of the things which "normal" people value are also valued by the handicapped. Mentally retarded children will be as delighted if they show progress, e.g. in learning to use words or even to read, as "normal" children will be. However, sometimes one does not know how individual handicapped people perceive and evaluate their situation and conditions for living. For instance, do they want to live in smaller or in bigger groups? These questions cannot be answered by others, they have to be answered by the handicapped themselves, from their point of view. Here more subjective quality-of-life research is needed. Other important questions for subjective quality-of-life research are: How do the handicapped experience their handicap? To what extent does their handicap and their being conscious of their handicap influence their satisfaction with life? Which special measures, activities etc. can improve their satisfaction with life? Such research is important, not only because it can contribute to the improvement of the situation of the handicapped, but also because it can help to demythologize all kinds of preconceptions about them. The handicapped are easily regarded as pitiful and less happy than the non-handicapped. Although the range of alternative life-options for most handicapped people will be limited, that does not automatically imply that they are less satisfied with their life. Research on handicapped people's satisfaction with life is important for refuting stigmatizing ideas. For sensitive non-handicapped people it is perhaps possible to imagine how it would be to live with a certain physical handicap, although I suspect that they in general will have too negative an idea of the handicapped person's satisfaction with life.

The knowledge that there is often only a loose connection between objective conditions and subjective evaluation has consequences for the forms and content of care, education, work and other activities for the handicapped. Although teachers, social workers, care-workers and others hope that everything they do will improve handicapped people's satisfaction with life, they should realize that satisfaction with life is often difficult to influence from the outside. Satisfaction with life much depends on idiosyncratic factors. I suppose that progress in — intellectual, social and emotional — development will correlate positively with satisfaction with life. But not in all cases. Progress in development is a clear goal. It is visible and measurable. Progress is also the best reward for the activities of parents and teachers. But what can be said about improving quality of life if a handicapped person does not show any progress in development? I presume that the quality of life of such a person can only be improved if one does know him or her very well.

5.2. (Non-)treatment decisions and euthanasia

Handicapped people can fall ill, as do the non-handicapped. If quality-of-life considerations are relevant to medical choices regarding the non-handicapped, there is no reason to suppose that they will not be relevant to decisions regarding the handicapped. A crucial question is, what is the role of the handicap in such decisions?

Every handicap is already a burden. In deciding about treatment for a handicapped person, we should realize that the burdens of a treatment may add to the burden of the handicap. Therefore a handicapped person's balance of the benefits and burdens of a treatment may differ from that of a non-handicapped person. If a handicapped person cannot decide for himself, others will have to infer what that balance may be. Of course, generalizations are dangerous. As I said, handicaps differ as to nature, cause and severity.

Like anyone else, a severely handicapped person can get a life-threatening disease. Sometimes his family and careworkers will be tempted, if they have to decide about treatment or non-treatment decisions, to use that situation as an opportunity to let him die, even if the prognosis concerning the success of the treatment is good. In such a situation it is imperative to make one's motives very clear. Is the desire to let him die caused by horror of the severity of the handicap? Is the motivating reason that one feels oneself hurt by the sight of his condition? Is one's own well-being at stake, or the well-being of the handicapped person? These are understandable reasons and motives. However, in my view, the well-being of the handicapped person should always be the guiding principle.

It is understandable and justified if parents and doctors decide not to treat a very retarded child with cancer chemotherapy if it is impossible to explain to the child why he has to suffer from the side-effects of such a treatment with no guarantee of substantial success. In such a case the fact of the child's handicap does constitute a relevant factor in deciding about whether or not to install a treatment. In many other cases it does not. If a mentally handicapped person would benefit from a hip replacement and if a non-handicapped person in the same situation would be entitled to get one, the handicapped person also should get one.

Decisions in life-threatening situations should not be used as an opportunity to end the life of a handicapped person whom one thinks would be better off dead. In my opinion the right moment to decide about a handicapped child's life and death is directly after its birth. Many defective newborn babies will die if they do not get an appropriate treatment. As I argued before, it is unavoidable that a decision as to whether to provide a treatment for such a child will be based upon considerations regarding its future subjective and objective quality of life or well-being. If one judges that the child has enough chances for a worthwhile existence, one thereby commits oneself to doing everything one can for the child's survival and well-being. Unless, of course, the prognosis turns out to be wrong and the child's condition becomes worse than expected.

Quality-of-life considerations are — and should continue to be — used, not only in medical decision-making, but also in other decisions regarding the handicapped. The fear of and rejection of such considerations stem from an unjust identification of activities to improve quality of life with eugenetics. In the United States, in the 1985 regulations of the Department of Health and Human Services which provide the guidelines required by the 1984 Amendment to the Child Abuse Prevention and Treatment Act, it is stated that it is illegitimate to base decisions regarding discontinuation or non-instalment of treatments to retarded or disabled persons on "subjective opinions about the future 'quality of life'" (*Federal Register*, 1985, 14880; quoted by Zaner [20], 77). The use of the term is associated with an ethic which justifies abortion-on-demand. In deciding about the treatment of handicapped people, it is imperative to include such not directly medical considerations as intensity of pain and suffering, degree of normal functioning and independence and possibility of having a satisfying life.

NOTES

[1] A. Abbey & F.M. Andrews [2] showed that, besides coping abilities psychological features as internal control and performance are important determinants of (subjective) quality of life.

[2] Research by Mc Neill et al. [14] has shown that not all patients in the same circumstances also prefer the same treatment option. In the case of lung cancer, many patients prefer radiotherapy over an operation, if the risk of dying after an operation is greater than 10%, although an operation offers the best chance for survival. Risk avoiding behavior is stronger among older than among younger patients.

[3] Helga Kuhse [11] has shown that proponents of a sanctity-of-life ethics cannot avoid using quality-of-life considerations. As I noted, the more moderate adherents are aware of that. They do not have objections against the use of — subjective or objective — quality-of-life considerations in deciding between treatment and non-treatment. For them, respect for life *forbids* any form of direct, *active killing*. They associate the use of quality-of-life considerations with the rejection of the distinction between letting die and killing.

BIBLIOGRAPHY

[1]   Aristotle, *The Ethics of Aristotle. The Nicomachean Ethics*. Translated by Thomson, J.A.K. Revised with Notes and Appendices by Tredenick, H., Introduction and Bibliography by Barnes, J.: 1955, Penguin Books, Harmondsworth.

[2]   Abbey, A. and Andrews, F.M.: 1985, 'Modelling the Psychological Determinants of Life Quality', *Journal of Social Indicators Research*, *16*, 1-34.

[3]   Calman, K.C.: 1984, 'Quality of Life in Cancer Patients — An Hypothesis', *Journal of Medical Ethics*, *10*, 124-127.

[4]   Campbell, A., Converse, P.E., and Rodgers, W.L.: 1976, *The Quality of American Life. Perceptions, Evaluations and Satisfactions*, Russell Sage Foundation, New York.

[5]   Cribb, A.: 1984, 'Quality of Life: A Response to K.C. Calman', *Journal of Medical Ethics*, *10*, 142-145.

[6]   Fromm, E.: 1955, *The Sane Society*, Rinehart, New York.

[7]   Fromm, E.: 1976, *To Have or to Be?*, Harper & Row, New York.

[8]   Galbraith, J.K.: 1958, *The Affluent Society*, Houghton Mifflin, London.

[9]   Galbraith, J.K.: 1967, *The New Industrial State*, Hamish Hamilton, London.

[10]   Karnofsky, D.A. and Burchenal, J.H.: 1949, 'The Clinical Evaluation of Chemotherapeutic Agents in Cancer', in MacLeod, L.C.M. (ed.), *Evaluation of Chemotherapeutic Agents*, Columbia University Press, New York.

[11]   Kuhse, H.: 1987, *The Sanctity-of-Life Doctrine in Medicine. A Critique*, Clarendon, Oxford.

[12]   Leenen, H.J.J.: 1985, 'Kwaliteit van leven, een bruikbaar begrip?', *Tijdschrift voor Gezondheidsrecht*, 152-156.

[13]   Marcuse, H.: 1964: *One-Dimensional Man. Studies in the Ideology of Advanced Industrial Societies*, Routledge & Kegan Paul, London.

[14]   McNeill, B.J., Weichselbaum, R., and Pauker, S.: 1978, 'Fallacy of the Five-Year Survival in Lung Cancer', *The New England Journal of Medicine*, 299, 1396-1401.

[15]   McNeill, B.J., Weichselbaum, R., and Pauker, S.: 1981, 'Speech and Survival', *The New England Journal of Medicine*, 305, 982-987.

[16]   McNeill, B.J., Pauker, S.G., Sox, H.C., and Tversky, A.: 1982, 'On the Elicitation of Preferences for Alternative Therapies', *The New England Journal of Medicine*, 306, 1259-1262.

[17]   Meadows, D.L. *et al.*: 1972, *The Limits to Growth. Report for the Club of Rome*, Earth Island, London.

[18]   Ordway, S.H.: 1953, *Resources and the American Dream. Including a Theory of the Limits to Growth*, The Ronald Press Comp., New York.

[19]   Osborn, H.F.: 1957, *The Limits of the Earth*, Faber & Faber, London.

[20]   Zaner, R.M.: 1986, 'Soundings from Uncertain Places. Difficult Pregnancies and Imperiled Infants' in Dokecki, P.R. and Zaner, R.M. (eds.), *Ethics of Dealing with Persons with Severe Handicaps. Toward a Research Agenda*, Paul Brookes, Baltimore/London, pp. 53-71.

SECTION III

MEASURING QUALITY OF LIFE IN HEALTH CARE

RAY FITZPATRICK AND GARY ALBRECHT

THE PLAUSIBILITY OF QUALITY-OF-LIFE MEASURES IN DIFFERENT DOMAINS OF HEALTH CARE

'Quality of life' is a key concept in the contemporary practice of medicine and delivery of health care. Numerous developments in modern medicine, such as psychotropic drugs, open heart surgery, and joint replacement surgery have the capacity to improve individuals' quality of life. Conversely, many therapies, such as renal dialysis and ostomy surgery for cancer, whilest increasing survival may impose costs in reducing patients' well-being. It would be impossible even to begin to sum up the costs and benefits of contemporary health care without resort to the notion of quality of life. The controversy associated with the emerging importance of the quality-of-life construct in health care is provoked by attempts to *define, assess or measure* the construct. Some critics even claim that quality-of-life constructs cannot and should not be measured. Their objections vary from arguments about the basic plausibility and validity of measurement systems [57], through to ethical and philosophical critiques regarding real-world applications of quality of life measurement systems [76]. The intensity of debate surrounding attempts to introduce measures of quality of life into health care suggests that measurement systems are widely adopted. In reality, in the diverse fields of research, clinical practice and resource allocation, resort to formal methods of quality of life assessment have been very few. A dispassionate observer would be inclined to conclude from the limited impact of measures to date that the various critiques of quality-of-life measurement are right in asserting that quality-of-life measures are less accurate and useful than their advocates suggest.

This paper starts from the present precarious and, at best partial acceptance of measures of quality of life in health care and raises the question of their potential for more widespread application. The main approach is to speculate about the ultimate legitimacy of quality-of-life measures in health care. By legitimacy is meant on the one hand basic acceptance of these quality-of-life measurement systems as plausible, valid and coherent. But also to be legitimate measurement systems need to fulfil particular functions in particular institutions and contexts in ways that are accepted by the social actors involved. This kind of legitimacy requires examining not just the content of measurement systems but the ways in which they are applied. The sequence

L. Nordenfelt (ed.), Concepts and Measurement of Quality of Life in Health Care, 201–227.

is therefore as follows. Firstly, we argue that quality-of-life measures are intended to be used in a diverse range of contexts and applications and speculation about the legitimacy of measurement systems cannot be carried out without reference to this diversity. Secondly, we examine the contents of measurement systems in terms of whether they do offer plausible and coherent approaches particularly to address problematic features of the construct of quality of life such as its inherently personal nature. Thirdly, and lastly, we discuss the acceptability of quality-of-life measures *as used* in their diverse contexts. We conclude that existing measurement systems have both internal coherence and validity and are likely to be widely accepted in certain contexts. In particular applications, however, especially resource allocation, they are likely to remain controversial and contested.

DIVERSITY OF APPLICATIONS

The first point to emphasize in evaluating the potential or actual legitimacy of quality-of-life measurement in health care is that a wide range of different fields of potential application have emerged. The plausibility of measurement systems may vary enormously between applications. Quality-of-life assessments may become accepted and unproblematic tools in one domain of health care whilst remaining precarious and contested in other fields. The following five applications are identified and briefly discussed: health needs assessment; clinical trials; evaluation research; individual patient care; resource allocation. In each application the use of quality-of-life measures is significantly different and the points of dispute and controversy vary.

Health needs assessment

It is increasingly argued that health care systems, instead of merely reproducing professionally determined patterns of care, need actively to identify and prioritize the health needs of the populations that they serve. An essential part of this new emphasis on health needs assessment is the use of conventional epidemiological techniques for the analysis of morbidity and mortality to identify needs for health services. However, there is abundant evidence that current epidemiological measures do not encompass dimensions of health that are salient for patients. The patient is concerned with the ways in which disease impinges on personal experience and ability to function. For this reason analysts suggest that recently developed health

status- or quality-of-life measures, many of which have been specifically designed for use in population surveys, are important means of providing more
patient-centered measures of health needs [42]. They have been used in this
spirit to assess levels of self-reported morbidity of particular geographical
populations [20], of populations covered by particular health care services
[6] and of particular social groups such as ethnic minorities [3] and the
unemployed [52].

The emphasis in all such studies is in providing evidence of health needs
as defined by lay populations rather than by medically determined criteria.
The methodology of survey research makes it possible to obtain more
perceptual and socially derived data about the impact of illness in populations. Such instruments have been shown to be predictive of mortality [89]
and of the volume and costs of health care associated with particular patient
groups [91]. Of course it remains the case that subjective dimensions of
health status and quality of life as reflected in such prevalence surveys are
determined in advance by professional expertise in instrument design rather
than emerging from lay inputs. However, there are established methods of
maximizing the contribution of non-experts in determining the contents of
measuring instruments (content validity).

It is only a small logical step to move from using quality-of-life assessments to identify health needs in populations to their use as screening
instruments for patients seeking health care. Health professionals are not
very sensitive to patients' disability, general well-being, or quality of life
and tend to under estimate many problems that patients experience in
relation to disease [82] [30] [75]. It is therefore argued that quality-of-life
instruments may be used to screen patients and the results given to their
doctors to increase their understanding of patients' problems [5]. Screening
instruments have been shown to be of value in improving detection of
psychiatric illness [45] and it may be hoped that similar improvements can
be obtained from quality of life instruments.

Clinical trials

Many medical interventions have as a main objective, not increased survival
but improvement in patients' well-being. Examples would include joint
replacement surgery which is intended to improve the patient's range of
movement and ability to lead a more normal life; pharmacological management of diseases such as asthma and epilepsy, in which the objective is to
improve the individual's overall sense of control over symptoms; psycho-

therapeutic or medical management of schizophrenia, depression or other psychiatric disorders where the objective is not only improved function in the patient but also for his or her family. The generic term for the range of patient-based measures used to assess outcomes of medical interventions in clinical trials has become 'quality of life' although the meaning of the term varies between trials. If used in the context of a randomized controlled trial with careful attention to usual requirements of trial design and careful attention to measurement issues, quality-of-life measures can provide invaluable evidence of the benefits or harms associated with a study intervention compared with alternatives such as placebo [18]. Such studies have demonstrated the differential effects on areas of quality of life of alternative drugs to control hypertension [19]. Clinical trials with quality-of-life outcomes have also specified benefits of drugs in the management of rheumatoid arthritis [9], advanced breast cancer [17] and coronary artery disease [65]. The use of quality-of-life outcomes in trials may also be extended to evaluate alternative methods of delivering care as in studies of the benefits of multi-disciplinary teams compared with conventional management of arthritis [2] or of home-delivered occupational therapy [38].

Evaluation research

In the same way as with clinical trials, quality-of-life assessments are also increasingly used in health services research and evaluation studies. The design of such studies are not quite so precise as with clinical trials. With a few exceptions [84], they tend to be observational studies rather than using randomisation, although usually employing designs which assess patients before and after treatment. Such studies employing quality-of-life outcomes have been used to examine impact of coronary artery bypass grafting [14], prostatectomy [31] or medical treatment of rheumatoid arthritis [91]. One particular problem with non-randomised observational studies is the doubt as to whether observed differences in quality-of-life scores in one treatment arm of a study may be due to pre-treatment differences between groups rather than effects of treatment. However with careful design and sufficient numbers recruited to the study, it is possible to overcome potential limitations of non-experimental design and to identify important costs and benefits of particular treatments [86]. The nature and scale of quality-of-life outcomes of alternative treatments can then be given as information to future patients to enable them to decide about treatments in the light of personal preferences [85]. As with applications in clinical trials the purpose of using quality of life data in this context is to provide fairly precise information about personal consequences of specific treatment options to facilitate more informed choice by both patients and clinicians between treatment options. In this area too the use of quality-of-life measures has not proved particularly controversial.

Clinical care

For the most part quality-of-life measures are not currently used in routine individual patient care. Their potential use in screening patient groups has already been discussed and most forms of use of quality-of-life information with individual patients are only a modest extension of the principle of screening. Advocates argue that clinicians ask patients to complete questionnaires so that the clinician has fuller information about the patient [87]. There is one more elaborate form of use of quality-of-life assessments which involves the formal eliciting of patients' preferences for alternative treatments and their associated risks and benefits in order that patient and doctor can jointly reach more informed and shared decisions about treatment [59]. Methods for formally eliciting patients' preferences have been discussed [53] but are probably considered too elaborate for widespread use at present. An alternative use of quality-of-life information in clinical care is the use of prior research data about outcomes of alternative treatments to assist a patient in reaching an informed decision [58]. The emphasis in this application is upon providing information that is relevant and useful to the individual patient to increase the extent to which he or she may play an informed role in decisions about treatment.

Resource allocation

The application of quality of life which has generated the most controversy has been that of resource allocation. Advocates of this application argue that, given that demands upon health services are considerable and resources limited, some more rational system of allocation is required than current tacit and informal methods. The particular method associated with quality-of-life measures involves using data regarding the costs of various health care interventions in conjunction with their respective health benefits expressed in terms of survival adjusted for the quality of life (so called "quality adjusted life years"). With this data, cost-utility analyses are performed and interventions which provide the greatest gain for a unit of health care expenditure should be favoured in terms of resources over interventions with poorer cost-benefit ratios. The principles behind such proposals are a form of utilitarianism, which is persuasive. Various ethical objections have been raised about the likely consequences. For example, the elderly would receive less health care because of the low scores for quality adjusted life years associated with many interventions for this age group. Other objections emphasize the opposing principle of equity contradicted by utilitarianism. These broader philosophical issues between resource allocation on the basis of cost-benefit data are beyond the scope of this chapter. Of great importance however are the controversies surrounding the concept of quality of life asssociated with this approach.

For the other four applications of quality-of-life measurement in health care distinguished above it is possible (or, indeed, according to some [18] highly desirable) to keep systems of quantification simple. In particular for purposes such as screening, clinical trials and evaluation studies, many approaches to quantifying quality of life use simple scales to assess dimensions of quality of life such as physical, social and emotional function. Instruments often use ordinal assumptions to express individuals' scores. Moreover for most purposes dimensions of quality of life can and should be kept separate in data analysis and interpretation. It is advantageous for either an individual patient or a group of patients in a trial to know specifically that, for example, mobility has been improved as a result of intervention whereas social and psychological well-being has deteriorated. Nothing is gained by attempting to aggregate or average such contradictory trends. In the field of resource allocation, it is regarded as essential to develop more complex assumptions about measurement, the very complexity of which has stimulated scepticism and indeed controversy.

The overarching need in applications in the field of resource allocation has been to assign single consistent quantitative expressions of the benefits or outcomes in terms of quality adjusted life years gained from interventions. This has required the derivation of values or "utilities" to be attached to varying health states. The method which has to date been most frequently used has been the Rosser Distress Disability Index [69]. This index assigns all possible illness states to a unique position on a matrix of possible combinations of distress and disability. Quantitative values are assigned to combinations of distress and disability by a psychometric method in which rating panels perform magnitude estimations. Thus whereas for most quality of life applications, the objective is adequately to describe an individual's scores across a range of relatively simple scales or dimensions (pain, mobility, etc), the requirement in this context is to be able to assign one (usually interval or ratio level) value to the individual's overall health state.

In addition to moral and ethical objections to the utilitarian approach to resource allocation associated with QALYs there has been a substantial and largely critical reaction to the methodology associated with the calculation of utilities just outlined. Objections have variously focused on apparent variability of utilities between methods; the problems of deciding whose views are most appropriate to include in judgements regarding utilities; questions regarding the extent of social consensus regarding utilities.

To summarize the arguments of this section, quality-of-life measures in the field of health care have been developed for a diverse range of applications. At one extreme they have been used to improve the awareness of health authorities and of individual health professionals of the extent and nature of health-related problems to be addressed. This has largely been received as a benign and beneficial endeavor which, at best, increases the extent to which health care systems respond appropriately to the health problems which they confront. There has been a similar lack of controversy surrounding applications in trials and evaluation studies. If anything, the only problems seen as facing this application are those of a purely methodological nature concerning improvements in the ability to measure quality of life accurately. It has only been in the field of resource allocation that heated debate is the norm. The reasons for this debate are two-fold. One controversy concerns the attempt to assign interval level quantification to all states of health-related quality of life. The second controversy concerns the use to which such global measures are put and the acceptability of utilitarian approaches to resource allocation. However, questions regarding the legitimacy of this particular application are not necessarily relevant when appraising the role of quality of life measures in other fields.

VALIDITY OF QUALITY-OF-LIFE ASSESSMENTS

A central determinant of whether quality-of-life assessments become an influential element in health care must be their perceived validity. To the extent that quality-of-life measures provide reasonably accurate assessments of what they purport to address, their legitimacy must be enhanced. As argued earlier, most of the critical debate concerning quality of life has focused on moral and political aspects of applications in resource allocation.

Where critical attention has been given to the question of validity of quality-of-life assessments, it has usually been devoted to the most controversial of all measurement systems, namely systems of quantifying utilities such as QALYs. Generally, such critiques attempt to cast doubt on the plausibility of assigning unitary numerical values to varying combinations of quality and length of life [49] [68]. Whilest understandable, this concentration of attention on QALYs has resulted in the validity of simpler forms of quality-of-life assessment, more commonly found in the other four fields of application, being neglected. Sometimes the plausibility of any form of quality-of-life assessment is questioned on the basis of the undoubted difficulties associated with QALYs in resource allocation [57] [48] [83]. Discussion of the validity of quality-of-life measures has therefore been based on a rather narrow and particularly controversial form of the genre.

The validity of quality-of-life measures in the field of health is rarely taken for granted or glossed over by those actively involved in development of measures and is generally regarded as one of the most essential of psychometric properties to be established before an instrument can be used for any serious purpose. The view can still be found expressed by those unfamiliar with the field that questionnaire or interview based data are inherently more vague, "soft" and inaccurate compared with "harder" medical data such as radiological, laboratory, or clinical information. One response to such scepticism is to demonstrate that much quality-of-life data can be just as firm in the sense of reproducible as traditional medical variables [27]. This aspect of quality-of-life instruments is usually established by examining test-retest reliability of instruments. Indeed it is increasingly clear that whilst medical data may be obtained via "hard" media such as x-ray film or blood test, the meaning or interpretation that is inevitably required of the raw data may be more variable between observers than quality-of-life assessments [22].

However even if the epistemological distinction between hard and soft data in the field of medical measurement is of doubtful value in appraising quality-of-life measures, it may still be the case that many discussions of validity in this field have to date appeared to minimise the scale of the problems involved in establishing that instruments may accurately assess quality of life. In some cases the impression is conveyed that validity is confirmed by establishing agreement between quality of life measures and conventional medical measures of disease severity [23] [54]. The idea that quality-of-life measures can be validated by establishing agreement with medical measures of disease is not very convincing. Because of evidence that patients may successfully adapt to even severe disease, and also because of the theoretical expectation that quality-of-life measures should provide data that add to what is known from conventional measures, more sophisticated discussions down-play the role of validation of quality-of-life measures by reference to conventional medical measures. Indeed lack of association between medical and quality-of-life measure may even be regarded as evidence of the validity of a quality-of-life instrument [8]. There is a strong sense that in validating quality-of-life instruments by reference to agreement with more established measures, one is contributing to a somewhat different activity, ie enhancing clinicians' familiarity with newer instruments by emphasizizing their relationships with more routinely used measures such as x-rays or blood tests. However, increasing the familiarity or interpretibility of quality-of-life measures is not tantamount to establishing their validity.

A number of more plausible methods may be used to demonstrate validity. One approach is regarded by many as rather informal and as involving judgement rather than science, i.e., *face or content validity* [1]. The question asked by this approach is whether items in an instrument such as a questionnaire appear to address the range of topics intended. This can be examined by involving the widest possible range of expertise and social backgrounds in the development of an instrument or by asking respondents whether relevant items have been omitted from pilot versions. The reservation of critics such as Mulkay and colleagues [57] is that items and dimensions of instruments reflect analysts' and observers' preoccupations. There is no formal way of completely disproving this observation which may in any case be reduced to the trivial truth that any instrument on any subject in any field of enquiry to some extent reflects the interests and agenda of its developers. However, although short of complete proof, every effort can be made to identify omissions of important dimensions from instruments.

The other method of validation used in this field is normally *construct validity*. With this approach a pattern of relationships is examined between quality of life measures and other variables theoretically considered relevant or related. The most formal and ambitious version of construct validation is convergent and divergent validity where overall patterns of extent of agreement with closer or more remote constructs are relevant. Studies of quality-of-life instruments which use variants of this approach are quite encouraging [12] [11].

However for sceptics of the field, approaches such as convergent and divergent validity, whilst examining formal statistical aspects of validity, do not address more fundamental reservations about the appropriateness of the content of typical instruments. They might argue that requirments to satisfy formal statistical tests of validity are so broad and general that it is difficult to determine precisely when specific instruments have passed or failed the criteria. To take an example, correlations of quality-of-life instruments with adjacent constructs such as severity of disease are expected to be neither too high nor too low. If they are too high, then the newer construct may be regarded as redundant because it adds no additional information about subjects. If correlations are too low, then by conventional statistical criteria, there is no evidence of construct validity. As a result this statistical requirement for construct validity — that correlations with related constructs be neither too high nor too low is rather vague and general so that formal statistical methods of validation may remain less than convincing.

Absent from current discussions of quality-of-life measures is a level of appraisal of validity which is intermediate between formal statistical criteria such as convergent and divergent validity and common-sense criteria such as intuitive inspection of the content of questionnaires (content validity). This intermediate level would assess the theoretical appropriateness of instruments to the specific task at hand. By this is meant that all instruments are developed with either implicit or (less commonly) explicit theoretical notions of such issues as illness, quality of life or function. By theory in this context is meant such matters as ideas about how illness affects daily life, notions of what constitutes disability and how such phenomena are to be assessed. An example of an explicit theoretical approach to quality of life would be the emphasis of the widely used Sickness Impact Profile on behavioral dysfuction [7]. An example of a more implicit theory embedded in a quality-of-life instrument would be the contents of the Health Assessment Questionnaire where respondents' scores are systematically elevated if they have such difficulty in performing any task that they require help from others (a dependency model) [34].

A simple example can be used to illustrate the argument. Two multi-dimensional health status instruments were used to assess patients with rheumatoid arthritis — the Arthritis Impact Measurement Scales (AIMS) [54] and the Nottingham Health Profile (NHP) [43]. Both instruments include a mobility scale and social scale to assess impact of illness on social relations. Patients completed the two instruments on two occasions three months apart [28]. On both occasions and also for changes between the two occasions, the two mobility scales produced reasonably consistent pictures of the sample of patients, whereas there was hardly any concordance between patients' scores for the two social scales. In this example the discrepancies between the two instruments can fairly easily be identified. The AIMS social scale focuses upon frequency of contact with social networks. The NHP social scale is more concerned with respondents' emotional responses to their social relations (perceptions of isolation and of being a burden). The two instruments draw on behavioral/structural notions of social relations in one case and a model of the emotional and supportive functions of social relations in the other [74] [59]. Both factors have been shown to affect well-being and are clearly potentially relevant dimensions of quality of life. However, it is clear that conceptually the two instruments do not contain synonymous constructs and in the case of the study of patients with arthritis different pictures of health-related quality of life were conveyed by the two apparently similar measures.

It may be argued that social relations are more likely to be associated with competing notions of what constitutes quality of life, whereas for other domains there is more agreement. The same study compared patients' scores for other dimensions of quality of life as reflected in different commonly used instruments all completed by patients at the same time [29]. Even for more "objective" and behavioral dimensions of instruments such as activities of daily living and role performance, substantial discrepancies emerged between instruments. Such results are not suprising. In subtle but pervasive ways different instruments contain different conceptions of disability; for example, some of the AIMS scales require respondents to resort to help in performing daily tasks in order to register for any degree of negative score. By contrast, on the Health Assessment Questionnaire although evidence of dependency is used to adjust scores, nevertheless respondents begin to score negatively on items as soon as they experience any degree of personal difficulty in carrying out daily tasks. Both instruments operate with a dependency model of quality of life, but one also emphasizes subjective distress and difficulty as a component.

If instruments contain within them tacit theories of reality, they may also contain alternative theories of how that reality is to be captured. Mor and Guadagnoli [56] argue that quality-of-life measures can be distinguished in terms of which of at least three different notions of how such measures are to be captured. The first group of measures emphasize reliance on "objective" indicators such as behaviors, that can be reported by the patient but also in principle verified by an observer. A second approach relies on subjective phenomena such as the experience of distress, difficulty, symptoms or other sensations. Thirdly, another group of instruments contain primarily judgements of a relativistic kind of how things are now compared either with how they been in the past or compared with what others might expect. This approach to quality of life is also particularly subjective since judgements are involved with often poorly specified anchors.

The thrust of this discussion is that methods of validating quality-of-life instruments are diverse and range from informal appraisal of the appropriateness of instruments through to elaborate statistical analysis. Methods of validation have focused excessively on convergence with medical methods and statistical criteria and have neglected attention to alternative assumptions behind different instruments. However, currently informal methods can be built on in the future more rigorously to establish appropriateness of instruments to the particular disease groups or interventions under examination.

Personal priorities

One of the strongest objections to quality of life measurements in the field of health care will remain that, even if the concept is of importance, and even if some degree of validity can be established for measurement of basic dimensions, it is an unavoidably personal construct which cannot be measured by methods that assume commonly shared values and meanings. Even if it is conceded as possible to identify relevant dimensions of quality of life, such as pain, psychological well-being, and social relationships, it is not possible to assign values regarding the personal significance to any individual of one domain compared to another. Thus Mulkay and colleagues argue that the quality-of-life assessments produced by current measurement systems are a social construct created by health economists' and others' measurement assumptions about priorities rather than reflecting patients' own perceptions [57]. Others question the degree of precision that can be claimed for general measurement systems for personal priorities. For example individuals will vary in their perception of the point at which pain, distress and dependency would make life so miserable that it is viewed as worse than death [48]. At a less emotive level, two individuals may both experience problems with immobility and pain, however *their degree of concern about the two problems may be quite different*. Many instruments attempt to remain neutral with regard to relative weight and priority assigned to different domains of quality of life by treating all domains as numerically equivalent, with an identical range of scores. However, even this apparently sensible and neutral response to the problem of priorities between areas of quality of life is actually based on the most controversial of all assumptions. Methods of quantification that assign equal numerical importance to different domains (e.g., Spitzer's QL Index which assigns equal value to five domains of quality of life for all respondents [77] in fact make the strongest of all assumptions that domains of quality of life are of equivalent importance.)

A number of different approaches have been adopted to address problems of personal priorities and preferences between dimensions of quality of life. One solution is to obtain general weightings obtained from panels' consensus judgements. This raises the broad question of whether there exists a social consensus about quality-of-life preferences that is elicited by such rating exercises. Despite some early studies indicating differences between income groups in preferences for different health states [71], generally the available evidence suggests that there is less disagreement about such valuations between social groups than might be expected [35]. Indeed, an international study found that panels from the United States and England came to very similar conclusions regarding values assigned to different health states [67]. Similarly, there is only moderate evidence of substantial differences betwen age groups in terms of health status or quality of life preferences [35].

Of greater concern is the evidence that individuals' own health status may make a difference to quality-of-life preferences. For example, patients on renal dialysis may rate this state much more favorably than do well populations [71]. Those with greater experience of health and illness such as doctors may also differ in their judgements from well individuals [69]. There is also evidence that patients' quality-of-life preferences may change in the course of treatment [64]. Concern about the validity of panels' values as scores for quality-of-life preferences also arises from the evidence that subtle shifts in the instructions of how panels are to obtain their judgements can alter results. For various reasons the use of consensus scores from panels of the well to assign values to different dimensions of illness for the sick seems unsatisfactory and various other solutions have recently been attempted.

A number of techniques have emerged which attempt to maximize the potential for individuals to express personal preferences with regard to quality of life domains — i.e., to state which domains are of particular importance in relation to health problems and health care. Thus women with metastatic breast cancer were asked to state which domains of quality of life were of particular importance to them [78]. Patients assessed quality-of-life priorities through two different techniques (Q Sort and linear analogue ratings). The results of the two methods were very similar. Reliability of preferences was examined by retesting preferences six weeks later. Consistency of preferences was quite high.

A number of other approaches have built on this basic idea that patients can identify more specific, personal dimensions of health status or quality of life that are of particular concern. For example, MacKenzie and colleagues developed a technique whereby medical and surgical patients identified specific problems that were of particular concern in three domains (physical activity, mental effort and emotions) [50]. Patients were interviewed by two different interviewers and problems selected by patients were found to be quite consistent across interviews. A similar approach was adopted by a group concerned to develop a more individualized approach to assessing quality of life in rheumatoid arthritis [81]. Patients were asked to identify up to five activities that were particularly affected by their disease. The personalized questionnaire was used as a baseline and patients were asked to assess improvements after an eight-week period in the areas they had identified. They also completed a more traditional health status questionnaire at baseline and at eight weeks. The personalized version appeared to be more sensitive to patients' overall judgment of change than was the conventionally formatted questionnaire. This individualized questionnaire was subsequently compared to two conventional, standardized format quality-of-life questionnaires in a randomized controlled trial to compare benefits of a particular drug methotrexate (a slow-acting anti-rheumatic drug) with placebo [80]. The individualized questionnaire was more sensitive to differences between the active drug and placebo than were either conventional quality-of-life measures or indeed various conventional rheumatological measures. Individualized quality-of-life measures have also been attempted in a study of total hip replacement surgery [63]. Patients were asked to nominate five areas of life they judged to be most important to their quality of life. Only rarely were patients unable to carry out this task. They were also asked to rate how well off they were with regard to selected areas and how important each selected area was to overall quality of life. Individualized quality-of-life scores were calculated for before and after hip-replacement surgery, together with scores from two well-established standard-format instruments. Compared with a control group, patients experienced significant improvements in both conventional as well as individualized quality-of-life scores.

Thus in a number of different ways efforts are being made to make quality-of-life measures more responsive to individual preferences and priorities. These efforts have in common the intention to meet the objection that quality of life is inherently personal and variable between individuals. Whereas conventional instruments either assign standard values and numerical weights across domains or attempt to distinguish salience of domains by using generalized weights from panels' rating exercises, more recent instruments attempt to build in maximal choice between domains to reflect personal priorities of respondents more convincingly .

Judgements of change

A related issue concerns the assessment of change in quality of life. For most uses of "quality of life" identified earlier, individuals' actual scores at a point in time are of secondary importance. In the context of evaluation studies, clinical trials and individual patient care, quality-of-life assessments are intended to be used dynamically — to detect changes over time. The way in which this tends to be done with quality-of-life measures of is to calculate the extent of change in individuals' scores between an administration of the relevant instrument before treatment and scores obtained at important time(s) after treatment. Again, a number of attempts have been made to involve the patient more personally and directly in assessing degree of change in addition to or instead of calculating change acores between administrations of quality-of-life instruments.

The patient can be more directly involved in judging whether treatments have improved their health status or quality of life by being directly asked "transition questions" (whether quality of life has improved, remain stable or deteriorated compared to a specified previous occasion). Transition scales were first developed to improve the sensitivity to change of simple ordinal scales used by clinicians. For example, the New York Heart Association scale assesses patients in terms a four point scale, unimpaired, or slightly, moderately or severely impaired [37]. Such scales may be accurate in assessing a patient's state at a point in time but may be too gross to detect changes over time. Also being ordinal in nature, mathematical subtraction of one score from another to calculate change scores is not strictly warranted. So, transition scales were developed to enable the clinician directly to assess the degree and direction of change in patient's state compared with a previous occasion and have been shown to be sensitive to drug treatment of angina in a controlled study [26].

The same principle can easily be extended to patients' judgements as incorporated in quality-of-life measures. Patients' own judgements of change in quality of life have been shown to produce plausible and sensitive results

in a number of studies [50] [80]. In a study of changes over time in patients
with rheumatoid arthritis, patients own transition judgements were shown to
be more sensitive to changes over time in quality of life judged important by
other criteria than were change scores produced by calcuating the differences
betwen scores obtained by static measures [92].

In both of the examples of recent developments in quality-of-life assessments discussed here (individualized instruments and transition judgements),
developments have been proposed that not only appear to improve the
methodological sensitivity of measures in practically important applications;
they also address more fundamental philosophical reservations regarding the
inherently personal and variable qualities of quality-of-life judgements. It is
important also to note that such developments are largely intended to
improve measurement in clinical trials and evaluation research, and do not
address the particular complexities associated with more ambitious global
indices required for cost-utility analyses for resource allocation.

LEGITIMACY OF DIFFERENT APPLICATIONS

Ultimately the legitimacy of quality-of-life measures will depend not only
upon their internal coherence and pausibility but also whether they are
perceived as useful and acceptable in specific contexts and applications.
There is some evidence available already that can be examined in order to
assess the ultimate acceptability of quality-of-life measures in health care.
This assessment will be carried out in terms of the five fields of application
used earlier.

In the context of *health needs assessment*, there has been considerable
activity using quality-of-life measures to obtain more patient-focused, non-
medicalized data regarding the full impact of health problems on individuals'
lives. As chronic and degenerative diseases become an increasingly central
focuses of health care activities, so the medical model of disease will need to
be supplemented. Quality-of-life measures are viewed as a central part of
this revision of perspectives [87].

However, initial enthusiasm may now be in the process of being replaced
by greater caution at least in the context of population surveys of health
needs. One very valuable use of quality-of-life-methodology in Great Britain
has been the Office of Population Censuses and Surveys' study of disability
[51]. A complex and sophisticated approach to measuring severity of
disability was adopted, and care taken in sampling procedures. The survey
has produced invaluable evidence about practical government-generated

questions about the prevalence of disability and associated financial, health, and social service needs. Such exercises demonstrate the value of well funded and well conducted national surveys. Much less may be gained from additional local conduct of such surveys for applied purposes of informing health policy. A survey of the disabled in South London used a form of the Sickness Impact Profile to examine the prevalence of disability [66]. It uncovered little unmet need via the instrument. In relation to the few gains to be obtained from local service-oriented uses, the costs of administering and processing longer and more sophisticated quality-of-life instruments are high. Although well conducted scientific surveys provide important information regarding social factors in health and illness, for more applied purposes the value of quality of life surveys may be less clear. While they provide evidence of a wide range of broader impacts of disease, quality-of-life instruments do not provide clear evidence of specific health needs that may identify specific changes to patterns of health care provision. Population surveys may produce little variability in results and little information that is precise enough to be useful [47]. To be of use to health services, surveys need to produce evidence of needs for particular medical or social interventions which might benefit the targeted individuals [32]. It is not clear that quality-of-life instruments have that potential. A current fashion in Great Britain for using such instruments to assess health needs has been described as "displacement activity that is professionally reassuring at a time of uncertainty, rather than a productive means of "health needs assessment" ([33], p. 257).

The ultimate practical value of using quality-of-life instruments to screen patients actually attending for health care is also unclear. One survey asked consulters at a number of general practices to complete a quality-of-life instrument prior to attendance and four weeks after consulting. The results did confirm evidence from other studies about relationships between age, sex, and health-related quality of life variables,but the investigators struggled to find specific evidence of the value of the information gained for health care providers [41]. There have been a few more rigorously designed studies established specifically to examine the value of having patients complete quality-of-life questionnaires. In randomized trials subjects' but not controls' quality-of-life scores have been given to clinicians [70] [46]. However changes as a result of doctors' being better informed, either in how patients in the study group were treated or in terms of various outcomes were very few. Such studies indicate the need to think more carefully about *how* such information is to be useful for health professionals. In general, clinicians in such studies find the information interesting and informative and do not indicate that quality-of-life assessment *per se* is a questionable activity in this context.

In the *clinical trials* context, quality-of-life assessments have proved particularly fruitful. Institutions such as the pharmaceutical industry are becoming very interested in this new range of outcome measures. Many drugs have as their main end-point improvements in patients' well-being rather than survival and favorable impact upon patients' quality of life is a new and accessible form of publicity that may revitalize or encourage doctors' prescribing. There is indeed now a substantial amount of enthusiasm for the use of quality-of-life measures in clinical trials in which traditional end-points need to be supplemented by patient-based measures [21] [73]. At present, a minority of clinical trials include quality-of-life outcomes, but it would seem that the main reason for omission of such measures is lack of familiarity with the approach, rather than hostility or doubts regarding validity [21] [13]. In this application, the problems of using quality-of-life measures are considered manageable whereas in the context of health needs assessment, fundamental questions about use and utility have emerged. The difference may largely be that clinical trials address very specific questions, with clearly controlled designs and abundant additional information about patients available so that fairly confident inferences about the meaning of quality-of-life scores in that environment can be made [18].

In the same way, there is an almost unrealistic level of optimism in some quarters for the use of quality-of-life measures in *evaluation (or health services) research* [25] [79]. It is expected that, appropriately used, such measures will give unique evidence of the impact of health care systems on patients' well-being [36]. Practicing clinicians echo this expressed need for quality-of-life measures to be developed to assess the outcomes of care [44]. It has to be recognized that quality-of-life measures are less useful in more routine audit of health services and are only likely to be feasible in research settings [40]. Nevertheless, in well-conducted experimental, quasi-experimental or longitudinal observational studies with adequate sample sizes, quality-of-life measures have a distinctive role to play when used alongside information about processes of care to address specific focused questions. As with clinical trials, problems in their use concern selection of appropriate measures and sensible interpretation of evidence of changes over time apparently brought about by interventions [29]. There are few fundamental conceptual, moral, or institutional barriers to their use.

In the case of *individual* patient care, there is little if any evidence of widespread use of quality-of-life measures. Reference has already been made to research trials to examine the benefits gained from their use in screening patients as part of routine care. One problem frequently cited is that quality-of-life instruments are too complex and time consuming to be routinely used in individual patient care. Thus, most recently, efforts have been made to develop much shorter instruments which whilst less sensitive to many aspects of patients' quality of life may nevertheless stand a better chance of being taken up by busy doctors [61]. Another problem that may act as a barrier to their wider use is that some instruments, especially more detailed ones such as the Sickness Impact Profile, produce scores that are not intuitively understandable. Clinicians cannot judge the significance of scores in the same way as more familiar laboratory tests [10]. Yet, again, we would argue that there are few fundamental barriers to wider acceptance of such instruments in clinical care other than practical ones. In particular there is clear evidence that patients are, for the most part, happy to complete quality-of-life measures and indeed regard the information contained in such questionnaires important for the doctor to know [60] [55].

The realm in which quality-of-life measures will continue to engender controversy is resource allocation. In both Great Britain and United States quality-of-life measures have become closely associated not only with research-based trials and evaluation studies but also with more immediate and practical efforts to change the ways in which health care is allocated. In Great Britain a number of health authorities are actively considering whether to use QALYs in their decisions about purchasing health services [16]. In the United States particularly intense debate has followed Oregon's plans to prioritize health care services under its Medicaid program with criticisms particularly focusing on the validity of rankings of different health services determined by benefits to survival and quality of life [24].

Attempts to utilize quality-of-life measures in this public context are likely to prove precarious and be greeted with widespread public scepticism for many reasons. Firstly, it should be clear from what has been said so far that to date no instrument exists that is capable of being valid and relevant for all possible diseases and interventions. Such an instrument would have to be both maximally relevant to the widest possible range of health problems and health care interventions, yet also be capable of generating the single composite (global) scores which are the primary requirement of cost-utility studies conducted for resource allocation. Definitions of health related quality of life would have markedly to expand to incorporate notions of reassurance, support, autonomy that are central to many health care activities. A recent observation on available methodologies by a health economist is pertinent: "As Ghandi said of British civilisation, QALYs would be a good thing" ([62], p. 1575). Not only do such instruments not exist, there is also a distinct lack of relevant data about both quality of life, but also about survival [15]. Health authorities have so far failed to find sufficient data to inform initial exercises in using QALY principles to allocate health care [4]. The alternative to the massive range of research studies required to determine outcomes of interventions by relevant measures is to resort to expert opinion which is the very mechanism of subjectivity and arbitrariness that QALY-type procedures are intended to overcome.

It is also clear that the moral and ethical underpinnings of QALY-type procedures are far from secure. Against the principle of utilitiarianism involved in allocating resources by cost-benefit principles, there are counterposed desires to maintain principles of equity and fairness and it is clear that methods to obtain acceptable trade-offs between equity and efficiency have not been identified [72]. Ethical reservations about principles of application inevitably "spill over" to influence judgements of underlying quality-of-life constructs.

There are other contradictions and tensions within the attempt to use quality of life measures in resource allocation that may threaten the legitimacy of the whole process. In applications such as the National Health Service and Oregon, quality-of-life measures are applied in explicit as opposed to implicit or hidden rationing. Policy makers argue that public acceptability of rationing will be high because the public has participated in the criteria and values to be used in allocation (via QALYs or their equivalent), and because decisions are transparent, and accessible rather than hidden behind individual clinical decisions as in the NHS case. However this is a gamble. The public attractiveness of democratic open decision making may be outweighed by making rationing starkly visible. Also, public involvement in determining values behind quality-of-life measures and the legitimacy that may follow from public participation has to be balanced by the inevitably arcane and complex way in which such commonsense judgements provided by the public have to be transformed into QALYs and ultimately into rankings of health care interventions that determine funding. Finally, the whole procedure of using cost benefit data based on QALYs or related methods is tainted by the broader political context in which such methods are applied. In Oregon, for example, such procedures are associated with socially divisive rationing of health care services for the poor alone. The use of

QALYs for resource allocation in the NHS will inevitably be tainted with the uncertain legitimacy of an internal competitive market for health care.

To bring together the arguments of this paper, in the context of clinical trials and evaluation research, there are a number of precise, focused, questions which can only be properly addressed by quality-of-life measures. The alternative is that such questions are not addressed at all. Methods exist that attempt both to maximize the confidence with which judgements about quality of life are made in such studies while also remaining substantially close to patients' own subjective judgements. The science and the use suggest a high degree of ultimate social legitimacy. By contrast, methods of deriving universally applicable measures that generate single global expressions of quality of life for the purposes of deriving quality adjusted life years or equivalent global expressions of health care benefits are barely explored and technically implausible. Moreover, their field of application provokes extensive controversy which further jeopardize efforts to produce technical solutions to resource allocation dilemmas. Intermediate in terms of plausibility are applications such as screening of patients and individual patient care, the logic and utility of which have not yet been clearly thought through. More generally, future discussions of quality of life in health care should be more specific and explicit about context and uses.

BIBLIOGRAPHY

[1]   Aaronson, N.K.: 1989, 'Quality of life assessment in clinical trials: Methodological issues', *Controlled Clinical Trials*, *10*, 196S-208S

[2]   Ahlmen, M., Sullivan, M., and Bjelle, A.: 1988, 'Team versus non-team outpatient care in rheumatoid arthritis: A comprehensive outcome evaluation including an overall health measure', *Arthritis and Rheumatism*, *31* (4), 471-79.

[3]   Ahmad, W.I., Kernohan, E.E., and Baker, M.R: 1989, 'Influence of ethnicity and unemployment on the perceived health of a sample of general practice attenders', *Community Medicine*, *11* 148-56.

[4]   Allen, D., Lee, R.H., and Lowson, K: 1989, 'The use of QALYs in health service planning', *International Journal of Health Planning Management*, *4*, 261-73.

[5]   Almy, T.: 1988, 'Comprehensive functional assessment of elderly patients', *Annals of Internal Medicine*, *109*, 70-72.

[6]   Anderson, J., Sullivan, F., and Usherwood, T.: 1990, 'The Medical Outcomes Study Instrument (MOSI) — use of a new health status measure in Britain', *Family Practice*, *7*, 205-18.

[7]   Bergner, M., Bobbitt, R.A., Pollard, W.E., Martin, D.P., and Gilson, B.S.: 1976, 'The Sickness Impact Profile: Validation of a Health Status Measure', *Medical Care, 14*, 57-67.

[8]   Bijlsma, J.W. Huiskes, C.J., Kraaimaat, F.W., Vanderveen, M.J., and Huber-Bruning, O: 1991, 'Relation between patients' own health assessment and clinical and laboratory findings in rheumatoid arthritis', *Journal of Rheumatology, 18*, 650-53.

[9]   Bombardier, C., Ware, J., and Russell, I.: 1986, 'Auranofin therapy and quality of life in patients with rheumatoid arthritis', *American Journal of Medicine, 81*, 565-78.

[10]   Brook, R. and Kamberg, C.: 1987; 'General health status measures and outcome measurement: a commentary on measuring functional status', *Journal of Chronic Disease, 40*, Suppl. 1, 131S-136S

[11]   Brooks, W.B., Jordan, J.S., Divine, G.W., Smith, K.S., and Neelon, F.A.: 1990, 'The impact of psychologic factors on measurement of functional status: assessment of the Sickness Impact Profile', *Medical Care, 28*, 793-804.

[12]   Brown, J., Kazis, L.E., Spitz, P., Gertman, P., Fries, J.,F., and Meenan, R.F.: 1984 'The dimensions of health outcomes: a cross validated examination of health status measurement', *American Journal of Public Health, 74*, 159-61.

[13]   Byrne, M.: 1992, 'Cancer chemotherapy and quality of life', *British Medical of Journal, 304*, 1523-24.

[14]   Caine, N., Harrison, S.C. Sharples, L.D., and Wallwork, J.: 1991, 'Prospective study of quality of life before and after coronary artery bypass grafting', *British Medical Journal, 302*, 511-16.

[15]   Carr-Hill, R.: 1989; 'Assumptions of the QALY procedure', *Social Science & Medicine, 29*, 469-77.

[16]   Carr-Hill, R., and Morris, J.: 1991, 'Current practice in obtaining the 'Q' in QALYs: a cautionary note', *British Medical Journal, 303*, 699-701.

[17]   Coates, A., Gebski, V., and Bishop, J.F. et al.: 1987, 'Improving the quality of life during chemotherapy for advanced breast cancer', *New England Journal of Medicine, 317*, 1490-95.

[18]   Cox, D.R., Fitzpatrick, R., Fletcher, A. Gore, S.M., Spiegelhalter, D.J., and Jones, D.: 1992, 'Quality of life assessment: can we keep it simple?', *Journal of the Royal Statistical Society/Series A General, 155*, 353-93.

[19]   Croog, S., Levine, S., Testa, M., Brown, B., Bulpitt, C., Jenkins, C., Klerman, G., and William, G.: 1986, 'The effects of antihypertensive therapy on the quality of life', *New England Journal of Medicine, 314*, 1657-63.

[20]   Curtis, S.: 1987, 'Self reported morbidity in London and Manchester: inter-urban and intra-urban variations', *Social Indicators Research, 19*, 255-72.

[21]   Deyo, R.: 1991, 'The quality of life, research and care', *Annals of Internal Medicine*, *114*, 695-97.

[22]   Deyo, R.A.: 1988, 'Measuring the quality of life of patients with rheumatoid arthritis', in S. Walker and R. Rosser (eds.), *Quality of Life: Assessment and Applications*, MTP Press, Leicester, pp. 205-22.

[23]   Deyo, R.A., Inui, T.S., Leininger, J.,D., and Overman, S.S.: 1983, 'Measuring functional outcomes in chronic disease. A comparison of traditional scales and a self-administered health status questionnaire in patients with rheumatoid arthritis', *Medical Care*, *21*, 180-92.

[24]   Dixon, J., and Welch, H.G.: 1991, 'Priority setting: lessons from Oregon', *Lancet*, *337*, 891-94.

[25]   Ellwood, P.: 1988; 'Shattuck lecture — outcomes management. A technology of patient experience', *New England Journal of Medicine*, *318*, 1549-56.

[26]   Feinstein, A., and Wells, C.: 1977, 'A new clinical taxonomy for rating change in functional activities of patients with angina pectoris', *American Heart Journal*, *93*, 172-82.

[27]   Feinstein, A.R.: 1983, 'An additional basic science for clinical medicine: IV. The development of clinimetrics', *Annals of of Internal Medicine*, *99*, 843-48.

[28]   Fitzpatrick, R., Ziebland, S., Jenkinson, C., Mowat, A., and Mowat, A.: 1991, 'The social dimension of health status measures in rheumatoid arthritis', *International Disability Studies*, *31*, 34-37.

[29]   Fitzpatrick, R., Ziebland, S., Jenkinson, C., Mowat, A., and Mowat, A.: 1992, 'Importance of sensitivity to change as a criterion for selecting health status measures', *Quality in Health Care*, *1*, 89-93.

[30]   Fossa, S., Aaronson, N., and Newling, D.: 1990, 'Quality of life and treatment of hormone resistant metastatic cancer.', *European Journal of Cancer*, *26*, 1133-36.

[31]   Fowler, F.J., Wennberg, J., Timothy, R.P., Barry, M.J., Mulley, A.G. and Hanley,E.: 1988, 'Symptom status and the quality of life following prostatectomy', *Journal of American Medical Association*, *259*, 3018-22.

[32]   Frankel, S.: (1991), 'Health needs, health-care requirements and the myth of infinite demand', *Lancet*, *337*, 1588-90.

[33]   Frankel, S.: 1992, 'The epidemiology of indications', *Journal of Epidemiology and Community Health*, *45*, 257-59.

[34]   Fries, J.F., Spitz, P.W. and Young, D.Y.: 1982, 'The dimensions of health outcome: The Health Assessment Questionnaire, Disability and Pain Scales', *Journal of Rheumatology*, *9*, 789-93.

[35]   Froberg, D.G., and Lane, R.L.: 1989, 'Methodology for measuring health-state preferences-III: population and context effects', *Journal of Clinical Epidemiology*, *42*, 585-92.

[36]   Greenfield, S.: 1989 'The state of outcome research: are we on target?', *New England Journal of Medicine*, *320*, 1142-43.

[37]   Harvey, R.M., Doyle, E.F., and Ellis, K., et al.: 1974, 'Major changes made by the criteria committee of the New York Heart Association', *Circulation*, *49*, 390-91.

[38]   Helewa, A., Goldsmith, C.H., Lee, P., Bombardier, C., Hanes, B., Smythe, H.A., and Tugwell,P.: 1991, 'Effects of occupational therapy home service on patients with rheumatoid arthritis', *Lancet*, *337*, 1453-56.

[39]   Henderson, S., Byrne, D., and Duncan Jones, P.: 1981, *Neurosis and the Social Environment*, Sydney, Academic Press.

[40]   Hopkins, A.: 1991 'Approaches to medical audit', *Journal of Epidemiology and Community Health*, *45*, 1-3.

[41]   Hopton, J.L., Porter, A.M. and Howie, J.G.: 1991, 'A measure of perceived health in evaluating general practice: the Nottingham Health Profile', *Family Practice*, *8*, 253-60.

[42]   Hunt, S., McEwen, J., and McKenna, S.: 1985, 'Measuring health status: a new tool for clinicians and epidemiologists', *Journal of the Royal College of General Practitioners*, *35*, 185-88.

[43]   Hunt, S.M., McEwen, J., and McKenna, S.P.: 1986, *Measuring Health Status* Croom Helm, London.

[44]   Hutchinson, A., and Fowler, P.,: 1992 'Outcome measures for primary care: what are the research priorities?', *British Journal of General Practice*, *42*, 227-31.

[45]   Johnstone, A., and Goldberg, D.P.: 1976, 'Psychiatric screening in general practice.', *Lancet*, I, 605-608.

[46]   Kazis, L.E., Callahan, L.F., Meenan, R.F., and Pincus, T.: 1990, 'Health status reports in the care of patients with rheumatoid arthritis.', *Journal of Clinical Epidemiology*, *43*, 1243-53.

[47]   Kind, P., and Carr-Hill, R.: 1987, 'The Nottingham Health Profile: A Useful Tool For Epidemiologists?', *Social Science & Medicine*, *25*, 905-10.

[48]   La Puma, J., and Lawlor, E.F.: 1990, 'Quality-adjusted life-years: ethical implications for physicians and policymakers', *Journal of the American Medical Association*, *263*, 2917-21.

[49]   Loomes, G., McKenzie, L.: 1989, 'The use of QALYs in health care decision-making', *Social Science & Medicine*, *28*, 299-308.

[50]   MacKenzie, C.R., Charlson, M.E., DiGiola, D., and Kelley, K.: 1986, 'A patient-specific measure of change in maximal function', *Archives of Internal Medicine*, *146*, 1325-29.

[51]   Martin, J., and Elliot, D.: 1992, 'Creating an overall measure of severity of disabi-
        lity for the Office of Population Censuses and Surveys Disability Survey', *Journal
        of the Royal Statistical Society/ Series A General, 155*, 121-40.

[52]   McKenna, S., and Payne, R.: 1989, 'Comparison of the General Health Question-
        naire and the Nottingham Health Profile in a study of unemployed and re-employed
        men', *Family Practice, 6*, 3-7.

[53]   McNeill, B., Weichselbaum, R., and Pauker, G.: 1978, 'Fallacy of the five year
        survival in cancer', *New England Journal of Medicine, 299*, 1397-1401.

[54]   Meenan, R.F.: 1982, 'The AIMS Approach to health status measurement: concep-
        tual background and measurement properties', *Journal of Rheumatology, 9*, 785-88.

[55]   Mercier, M., Schraub, S., Bransfield, D., and Fournier, J.: 1992, 'Patient
        acceptance and differential perceptions of quality of life measures in a French
        oncology setting', *Quality of Life Research, 1*, 53-61.

[56]   Mor, V., and Guadagnoli, E.: 1988, 'Quality of life measurement: a psychometric
        tower of Babel.' *Journal of Clinical Epidemiology, 41*, 1055-8.

[57]   Mulkay, M.J., Ashmore, M., and Pinch, T.J.: 1987, 'Measuring the quality of life:
        a sociological invention concerning the application of economics to health care',
        Sociology, 21, 541-64.

[58]   Mulley, A.G.: 1990, 'Medical decision making and practice variation', in T.
        Andersen and G.Mooney (eds.), *The Challenges of Medical Practice Variations*,
        MacMillan, London, pp. 59-75.

[59]   Mulley, A.G.: 1989, 'Assessing patients' utilities: can the ends justify the means?',
        Medical Care, 27, S269-81.

[60]   Nelson, E.C., and Berwick, D.M.: 1989, 'The measurement of health status in
        clinical practice', *Medical Care, 27*, S77-S90.

[61]   Nelson, E.C., Landgraf, J.M., Hays, R.D., Wasson, J.H., and Kirk J.W.: 1990,
        'The functional status of patients. How can it be measured in physicians' offices?',
        Medical Care, 28, 1111-26.

[62]   Normand, C.,: 'Economics, health and the economics of health', *British Medical
        Journal, 303*, 1572-77.

[63]   O'Boyle, C.,A., McGee, H., Hickey, A., O'Malley, and K., Joyce, C.R.: 1992,
        'Individual quality of life in patients udergoing hip replacement', *Lancet, 339*,
        1088-91.

[64]   O'Connor, A.M., Boyd, N.F., Warde, P., Stolbach, L., and Till, J.E.: 1987,
        'Eliciting preferences for alternative drug therapies in oncology: influence of
        treatment outcome description, elicitation technique and treatment experience on
        preferences', *Journal of Chronic Disease, 40*, 811-18.

[65]   Olsson, G., Lubsen, J., van Es G.A., and Rehnqvist, N.: 1986, 'Quality of life after myocardial infarction: effect of long term metoprolol on mortality and morbidity,' *British Medical Journal, 292*, 1491-93.

[66]   Patrick, D., and Peach, H.: 1989, *Disablement in the Community*, Oxford University Press, Oxford.

[67]   Patrick, D.L., Sittampalam, Y., Somerville, S.M., Carter, W.B., and Bergner, M:. 1985; 'A cross cultural comparison of health status values', *American Journal of Public Health, 75*, 1402-1407.

[68]   Rawles, J.: 1989, 'Castigating QALYs', *Journal of Medical Ethics, 15*, 143-47.

[69]   Rosser, R., and Kind, P.: 1978, 'A scale of valuations of states of illness: is there a social consensus?', *International Journal of Epidemiology, 7*, 347-58.

[70]   Rubinstein, L.V., Calkins, D.R., Young, R.T., Cleary, P., Fink, A., Kosecoff, J. et al.: 1989, 'Improving patient function: a randomized trial of functional disability screening', *Annals of Internal Medicine, 111*, 836-42.

[71]   Sackett, D.L., and Torrance, G.W.: 1978, 'The utility of different health states as perceived by the general public', *Journal of Chronic Disease, 31*, 697-704.

[72]   Schmidt, V.: 1991, 'Some equity-efficiency trade-offs in the provision of scarce goods: the case of lifesaving medical resources', Working paper, Centre for Social Policy Research, University of Bremen, Bremen.

[73]   Schumacher, M., Olschewski, M., and Schulgen, G.: 1991, 'Assessment of quality of life in clinical trials', *Statistics in Medicine, 10*, 1915-30.

[74]   Seeman, T., and Berkman, L.: 1988 Structural characteristics of social networks and their relationship with social support in the elderly: who provides support? *Social Science & Medicine, 26*, 310-37.

[75]   Slevin, M.L., Plant, H., Lynch, D., Drinkwater, J., and Gregory, W.M.: 1988, 'Who should measure quality of life, the doctor or the patient?', *British Journal of Cancer, 57*, 109-12.

[76]   Smith, A.: 1987, 'Qualms about QALYs', *Lancet*, I, 1134-36.

[77]   Spitzer, W.O., Dobson, A.J., Hall, J., Chesterman, E., Levi, J., Shepherd, R. *et al.*: 1981, 'Measuring the quality of life of cancer patients: a concise QL Index for use by physicians', *Journal of Chronic Disease, 34*, 585-97.

[78]   Sutherland, H.J., Lockwood, G.A., and Boyd,N.F.: 1990, 'Ratings of the importance of quality of life variables: therapeutic implications for patients with metastatic breast cancer', *Journal of Clinical Epidemiology, 43*, 661-66.

[79]   Tarlov, A.R., Ware, J.E.,, Greenfield, S., Nelson, E., Perrin, E., and Zubkoff, M.: 1989 The Medical Outcomes Study. 'An application of methods for monitoring the results of medical care', *Journal of American Medical Association, 262*, 925-30.

[80]   Tugwell, P., Bombardier, C., Buchanan, W.W., Goldsmith, C., Grace, E., Bennett, K.J., Williams, J.H., *et al.*: 1990, 'Methotrexate in rheumatoid arthritis;

impact on quality of life assessed by traditional standard item and individualised patient preference health status questionnaires', *Archives of Internal Medicine, 150*, 59-62.

[81]   Tugwell, P., Bombardier, C., Buchanan, W.W., Goldsmith, C.H., Grace, E., and Hanna, B.: 1987, 'The MACTAR Patient Preference Disability Questionnaire. An individualized functional priority approach for assessing improvement in physical disability in clinical trials in rheumatoid arthritis', *Journal of Rheumatology, 14*, 446-51.

[82]   Uhlmann, R.F., and Pearlman, R.A.: 1991, 'Perceived quality of life and preferences for life-sustaining treatment in older adults', *Archives of Internal Medicine, 151*, 495-97.

[83]   Wade, D.T.: 1991, 'The 'Q' in QALYs', *British Medical Journal, 303*, 1136-37.

[84]   Ware, J., Brook, R.H., Rogers, W.H., Keeler, E., Ross Davies, A., Sherbourne, C.D. et al.: 1986, 'Comparison of health outcomes at a health maintenance organisation with those of fee-for-service care', *Lancet*, I, 1017-22.

[85]   Wennberg, J.E.: 1990, 'On the need for outcomes research and the prospects for the evaluative clinical sciences', in T. Andersen and G. Mooney (eds.), *The Challenges of Medical Practice Variations*. Macmillan, London, pp. 158-73.

[86]   Wennberg, J.E., Mulley, A., and Hanley, D.: 1988, 'An assessment of prostatectomy for benign urinary tract obstruction', *Journal of American Medical Association, 259*, 3027-30.

[87]   Wilkin, D., Hallam, L., and Dogett, M.: 1992, *Measures of Need and Outcome for Primary Health Care*, Oxford University Press, Oxford.

[88]   Wolfe, F., and Pincus, T.: 1991, 'Standard self-report questionnaires in routine clinical and research practice — an opportunity for patients and rheumatologists', *Journal of Rheumatology, 18*, 643-46.

[89]   Wolfe, F., Cathey, M.A.: 1991, 'The assessment and prediction of functional disability in rheumatoid arthritis', *Journal of Rheumatology, 18*, 1298-1306.

[90]   Wolfe, F., Hawley, D.J., and Cathey, M.A.: 1991, 'Clinical and health status measures over time: prognosis and outcome assessment in rheumatoid arthritis', *Journal of Rheumatology, 18*, 1290-97.

[91]   Wolfe, F., Kleinheksel, S.M., Cathay, M.A., Hawley, D.J., Spitz, P.W., and Fries, J.F.: 1988, 'The clinical value of the Stanford Health Assessment Questionnaire Functional Disability Index in patients with rheumatoid arthritis', *Journal of Rheumatology, 15*, 1480-48.

[92]   Ziebland, S., Fitzpatrick, R., Jenkinson, C., Mowat, A., Mowat, A., (forthcoming). 'A comparison of two approaches to measuring change in health status in rheumatoid arthritis: the Health Assessment Questionnaire (HAQ) and Modified HAQ' (in press).

ANALYZING CHANGES IN HEALTH-RELATED
QUALITY OF LIFE

The notion 'quality of life' (QL) refers to how well an individual lives his or her life. This can be specified in terms of, e.g., the individual's emotional response to his or her situation [5] or according to some ethical standard for how to live the "good life". In both cases the QL of an individual depends on a number of circumstances such as education, social standard, etc. The more specific notion, health-related quality of life (HQL), has been defined in the following way by Patrick & Ericsson [6] "Health-related quality of life is the value assigned to duration of life as modified by the impairments, functional states, perceptions, and social opportunities that are influenced by disease, injury, treatment, or policy." Even though it is important to specify what "as modified by" means, this definition indicates that HQL is a more narrow notion than QL, *viz.* the QL affected by health and not the the QL affected or determined by, for instance, education. For our purposes these conceptual remarks are sufficient, as the main objective is to present a very practically oriented analysis.

HQL is the efficiency measure that has come to play an increasingly prominent role in the health-care sector. This has affected related disciplines: health economics, medical ethics, public health and social medicine, pharmacoeconomics, etc. HQL has become part of those disciplines.

There are a number of difficulties which can easily cause problems when measuring QL, i.e., anything from the definition of the concept and constructing a questionnaire to the choice of method for calculations and analysis of the results [1]. If these difficulties are not observed and resolved, studies may not lead to any useful result. We have also gained access to a series of general instruments for measuring HQL, such as MOS SF-36 [12] and EuroQol [10], only to mention two which have become more and more common.

However, this does not mean that studies on HQL are performed without problems. A good final result requires high competence on the part of the buyer of the study as well as on the part of those responsible for designing and performing the study. The development of methodology is making steady progress and in order to generate as much relevant information as possible with as few questions as possible, it is not sufficient to use only one

L. Nordenfelt (ed.), Concepts and Measurement of Quality of Life in Health Care, 229–240.
© 1994 *Kluwer Academic Publishers. Printed in the Netherlands.*

general questionnaire. Disease-specific and preferably also tailor-made questionnaires (according to the purpose of the study) are necessary in most cases [2].

There are many different definitions of the concept of HQL in the medical research field. One explanation is perhaps that the concept occurs in several different disciplines: philosophy, medicine, psychology, and economics.

Amartya Sen, who is active within economics as well as in philosophy, has put forward the distinction between individual features and activities [9]. It is important to explicitly specify the relation between these two concepts. The starting-point is that an individual has a high HQL when he or she can perform activities which he or she wants to perform. So, according to Sen, the possibility of *choosing* is important in order to achieve a high HQL. It is then possible to choose, as we do, to associate certain physical and psychological features with the concept of health.

If my physical/bodily structure is extremely weak but I have an exceptionally beautiful soprano voice, I ought not try to become a professional boxer if I seek high HQL. Instead, my choice of activities should be directed at training my voice and then at creating an opportunity to sing soprano in the most suitable way. If I should contract laryngitis, the effect of this would be worse for me than for a non-singing professional boxer who contracts laryngitis. There is no way of making a generally valid valuation of such a laryngitis.

What makes it even more complicated to attach a once-and-for-always valid weight to a health state is the fact that people may choose different activities given the same personal features. These are central issues to be considered when evaluating the other than purely medical effects of health care efforts. It is not only in a comparison of the same health states between different individuals that these individuals' valuations differ.

TABLE I. Appraisal of the HQL of patients treated with hypotensive drugs

Appraiser	Improved	Unchanged	Worse
Physician	100%	0	0
Patient	48%	43%	9%
Relative	1%	0	99%

(Source: Jachuck S.J. *et al.* [4], pp. 103-105.)

A study by Jachuck and colleagues shows that different individuals judge one individual's HQL differently [4]. In the study the physician, the patient and the patient's relatives gave their opinion as to whether the patient's HQL had increased, decreased or remained the same after treatment for hypertension (see Table 1). All three groups had different opinions. The physicians were of the opinion that all the patients' HQL had increased, while the relatives were of the opposite opinion, i.e., that almost all the patients' HQL had decreased. Most of the patients considered that they had achieved improved HQL or that it was unchanged. Only a little less than a tenth considered that their HQL had decreased.

There are at least two comments to be made as to the result. One is that different individuals judge one individual's health state differently. The other is that it is possible that different individuals assign different meanings to the concept HQL. Hence, different indicators are also chosen here for the characteristics of an individual's HQL. It seems reasonable to assume that the physician considers the blood pressure to be an indicator for HQL, whilst the patient simply pays attention to how he or she is feeling and whilst relatives perhaps primarily notice changes in the patient's mood affecting themselves. Individuals not only make different valuations of their own health state in comparison with somebody else in the same state but also value the health state of another individual differently than that individual himself/herself does.

Hence it is obvious that individuals value health states differently. In addition, they change their valuations over time. A sixteen-year-old youth values a health state differently than an eighty-year-old individual. In all health indexes and quality-of-life questionnaires, the health states included in the questionnaire are weighted in relation to one another. Traditional methods for weighting are based on the assumption that there is *one* fixed value assigned to each health state which applies to *all* individuals. This means that this health state must be independent both of contextual variables (personal preferences) not included in the questionnaire and of the respondent's age. Consequently, it is evident that a great deal of information is lost with such methods.

There are several different weighting methods [11]. They are all based on the assumption that different health states are weighted in relation to one another. The most frequently used methods for valuation are category scaling (CS), standard gamble (SG), and time trade-off (TTO). What separates them is that CS relates the values of *being in* different health states to a thermometer, SG values the *risk* of ending up in a certain health state and

TTO values the *time* that an individual can imagine being in a certain health state. Often, a mixture of these methods is used to allot weights to health indexes and quality-of-life questionnaires. The methods are based more or less on a theoretical ground and answer different questions.

Several scientific studies have confirmed that the different methods yield different results. In one study, the three methods were tested and compared with one another. Three different stages of angina pectoris were evaluated: none, moderate, and severe [7]. No stage of angina was allotted the value of 100. It was evident that the choice of method clearly influenced the relative valuation of the health state.

TABLE II. Mean values for different stages of angina pectoris, given three different methods for valuation.

Methods for valuation	No angina	Moderate angina	Severe angina
Category scaling	100	71.8	35.4
Standard gamble	100	90.3	70.7
Time trade-off	100	83.2	53.3

(Source: Read *et al.* [7], pp. 316-29.)

What we so far have discussed supports two relatively uncontroversial statements which may be expressed in the following way.

(1) Individuals value health states differently;
(2) Different weighting methods give different values to the same health state

The obvious way to handle these obstacles is to let each individual value relevant health states or relevant changes in health himself/herself. However, this presupposes a clear definition of HQL as well as a feasible method of calculation. The aim of this paper is to demonstrate our suggestion for dealing with these topics.

The success of the treatment of those topics does not only mean that it is possible to design new studies of health-related HQL, but also that it is possible to some extent to re-analyze the results of old studies. The basic idea is that it is important to specify the relation between personal features on the one hand, and the possibility of performing activities given those features on the other. Suppose both my legs have been amputated and that I am suffering from asthma. If I then answer a quality-of-life-questionnaire, and to a question about my physical mobility answer that my treatment for asthma has not improved my capability of climbing stairs, this is, with traditional methodology, interpreted as if the treatment did not have any effect on this. The conclusion is correct but uninteresting since no asthma medicine in the world could make my amputated legs grow again. Consequently, it is important to consider individual features in order to avoid misinterpretations.

In what way is the information on a patient's or on a group of patients' HQL of interest? Within the field of health care, producers and financiers as well as the patients themselves are interested in HQL as a result measure together with the medical quality of the outcome (mortality, morbidity, and recurrence) and the cost. Within the pharmaceutical industry there is of course a marketing interest, directed towards both consumers and pre-scribers/-physicians. The drug consumers' opinion of the total effect of a drug — its medical effect, side-effects, and the extent to which the drug affects their possibility of performing activities — is, of course, in itself interesting. However, this is also of great help in the marketing of the product as well as being part of the long-range strategic planning of a company.

SOME CONCEPTUAL FUNDAMENTS

Our theoretical base is multidisciplinary. In our sense of the notion high HQL presupposes that an individual *is able to* perform activities which he/she *wants to* do. In addition, he or she ought to have the right disposition for the activity in question. Therefore a certain degree of self-knowledge is necessary so that a person can choose the activities which will render him or her satisfaction and happiness, i.e. a high HQL [2]. When constructing an HQL index, it is important to observe the following central distinctions:

A.   Individual features;

B.   activities which an individual is able to perform, given his/her qualifications;

C.   activities which the individual actually chooses or would choose to perform, if he/she were made to choose; and

D.   the individual's own set of values which are the basis for the choices he/she will make.

Individual features which may also be called personal qualifications and re-sources ("personal resources" [2] or "personal features of the individual" [9]) comprise all the powers which an individual can summon up in order to carry out various activities. These qualities, qualifications, resources or fea-tures may be further specified. Hereafter, we will use the term "feature". A specified, full account of all the features of a person is too difficult to handle, as it would be very extensive. In order to avoid this, one can stipu-late which types of qualities to study. Here below, we will confine ourselves to health-related qualities. A complete specification of these would also be too comprehensive, however, so further definitions are justified. The indi-vidual's illness or injury and the treatments that may be necessary will de-cide the choice of features of interest.

Regarding angina pectoris, for example, it is important to specify the pain and anxiety connected with the disease itself. As to the treatment, it is vital to specify headache, for instance, because it is an undesirable effect of the treatment with nitrate plaster. Here, the starting-point should be to spec-ify the features and activities which can be expected to be affected by injury, illness or treatment.

A few examples will illustrate the connections between the four dimen-sions. The letters indicated refer to the above distinctions. If one has long legs and a good absorption of oxygen, one has also good physical features (see A above) for becoming a fast runner. If one has a stable mind as well and is stubborn, one has the right psychological disposition to also be able to achieve a good time in long-distance races (see B above). A fractured leg as well as a bad mental shock or angina pectoris can change these features (A), and all this also influences what activities a person is able to perform (B).

Individuals have different preferences (see D above) and choose different activities (see C above). For instance, some choose to play tennis, others to do housework or to cultivate their social lives. The activities which a person is able to perform (B) and also chooses to perform (C) are affected by factors like age and sex but also by health and access to health care services, education and information. To some, a fractured leg may be very serious, while it is of less importance to others. For a weight lifter it is more important to have both his arms intact than it is for an opera singer, just as it is more important for an opera singer than for a weight lifter to have fully working vocal chords. The importance of a changed state of health is dependent on what activities the individual prefers [2]. It is not obvious how a changed state of health will affect a particular individual.

An illness or an injury changes an individual's features more or less. The same goes for treatments. If the changed features (A) are important as regards the activities chosen by the individual (C) and perhaps also reduce the range of possible activities that the individual is able choose (B), this will affect the individual's HQL. However, in *what way* this affects the individual's HQL depends on the individual's own set of values (D) and how well those values reflect the "conceptual limitations" of the concept of HQL. (Those parts are not considered in this paper.) The expression "measuring changes in HQL" only involves A, B and C in this paper. This implies that we (in this paper) assume that the possibility of performing more activities rather than fewer activities is per se to be preferred. So the individual's own *valuation* of those parts is not included in the analysis presented here. The reason for this is practical: we do not have any information of this kind. As far as we can judge there is no limitation in this respect to the method presented below.

In the construction of an HQL index, it is important that the applied index method allows different choices of activities, that the same combinations of activities do not have to have the same value to all individuals, that different combinations of activities may have the same value and that individuals may differ as to how successfully they achieve their maximum level of activity. It is also vital to allow individuals to content themselves with less than their maximum level of activity.

ANALYZING DATA FROM HIP JOINT REPLACEMENTS

In this section we will illustrate a new approach for measuring changes in HQL. We will not discuss whether or not the variables used to describe health states and activities are the most relevant ones. Nor is it possible to present a detailed analysis of the results here. We have used data generated in a study of hip-replacements carried out at two clinics in Sweden [3].

Persons with a badly functioning hip joint have a reduced health state in terms of mobility, ability to sleep, etc. Hence HQL may vary between persons with hip joint problems and those with a better-functioning hip joint, as regards the activities that can be performed. So a hip joint replacement can change the HQL for the individual. However, it should be noted that a hip joint replacement *does not necessarily* lead to increased activity and improved HQL [2]. In this example we will observe (A) individual features, (B) activities which he/she is able to perform, and (C) activities which he/she has chosen.

The data have been gathered from questionnaires completed by 67 patients of an average age of 70 who have been operated on for hip joint replacement. The patients were asked about health-related features (A) and degree of activity (B and C) before the operation and six months after the operation. One index was constructed for the health-related features, the so-called Malmquist Health Index (MHI), and another index for the activities, the so-called Malmquist Quality of Life Index (MQLI). In this study we leave out the theoretical part of the method for calculating the Malmquist indices. See [8] for a presentation of the index constructions and the calculations using linear programming. The MHI measures quantitative changes of the individual's health. The MQLI measures changes in the quality of life of an individual, where "quality of life" refers to the ability of the individual to transform health features into activities. The following variables have been observed in the two indexes:

Health-Related Features

*   Physical activeness
*   Sleep and rest
*   Power of concentration
*   Mental balance

Activities

*   Personal hygiene
*   Housework
*   Social interaction
*   Leisure time and recreation

Every variable contains between 7 and 23 different questions which can be answered by "yes" or "no". Below, we will describe an alternative index approach based on MHI and MQLI for the measurement of changes in health states, where health state refers to the above-mentioned health-related variables.

MHI measures individual changes in health-related features. MHI has certain important characteristics:

(i)   MHI does not demand weights fixed in advance in order to aggregate health-related features;

(ii)   MHI allows for individual-based weights.

MHI measures *changes* in health-related features. Activities, however, are not excluded from MHI, so health-related features are always linked to a certain level of activity. So if a feature is changed, it will influence which activities can be performed. An activity also presupposes certain features. Table 1 shows the MHI results for 67 patients and the average for the whole group of patients.

MHI shows that the health-related features were improved by, on average, 21 percent. There were great variations between the individuals, with a maximum change of plus 156 percent and a minimum change of minus 20 percent. The cases of negative change may be explained by complications which had set in for some patients. Six months may also be too short a time between the two measurements. MHI shows a high average change and a great variation. This is explained by the fact that MHI considers *individual conditions.*

The next step of our example illustrates individual changes in HQL, where HQL is related to activities. MQLI measures changes regarding a person's ability to succeed in converting his/her features into activities, in this case after a hip joint replacement. MQLI also takes into account changes in the individual features. Changes in the activity index are here interpreted as changes in HQL. Based on the answers from all the 67 patients before an operation, an individual-based reference frame is established which can be said to show the maximum activity level possible for the individual, given his/her features. Correspondingly, a reference frame is fixed for the point of time six months after the operation. MQLI is affected if:

(a)   the reference frame is moved, i.e.; if the individual's highest activity level possible has increased/decreased;

(b)   the individual has come closer to /moved away from/ what is his/her highest activity level possible.

STEFAN BJÖRK AND PONTUS ROOS

On average the results show that the 67 individuals improved their HQL by 2 percent in terms of increased individual ability to convert their features into activities (Table 3). From the same table, it can be seen that the change in HQL differed between individuals. Individual 1 improved his HQL by 12 percent while individual 3 improved his by 2 percent. For individual 2, the results do not show any change in HQL. Most of the individuals (51) improved their HQL. For 9 individuals, the results show a decrease in HQL. In the calculations of MHI and MQLI, linear programming approaches were applied [8].

TABLE III. Change in HQL following hip joint replacement according to MHI and MQLI for 67 persons. Percentage changes.

Person	MHI	Change MQLI	Person	MHI	Change MQLI	Person	MHI	Change MQLI
1	+25	+12	26	+29	-2	51	+14	0
2	+19	0	27	-10	-6	52	+38	+3
3	+1	+2	28	+67	+1	53	+15	+1
4	+24	+4	29	+36	+9	54	+23	+1
5	+26	+7	30	+22	+2	55	+156	+2
6	+7	+2	31	+22	+4	+56	+6	+2
7	+32	+2	32	+38	+11	57	+11	+3
8	+24	+2	33	0	+3	58	+10	+9
9	+29	+3	34	+9	0	59	+16	0
10	+5	0	35	+17	+1	60	-20	-17
11	+34	+10	36	+16	+10	61	+17	+2
12	+7	-3	37	+58	+2	62	+21	+2
13	+18	0	38	+16	0	63	+45	+3
14	+25	+3	39	-2	+20	64	+18	+2
15	+85	-3	40	+11	+2	65	+26	+4
16	-7	-2	41	-7	+6	66	+13	+1
17	+23	+3	42	-2	+2	67	+13	+2
18	-3	+15	43	+22	+2			
19	+28	+9	44	+82	-9			
20	+12	+4	45	+26	+1			
21	+24	+1	46	-7	+16			
22	-1	+2	47	+105	-4			
23	+11	+1	48	+13	+6			
24	+13	+2	49	-8	+1			
25	+5	+2	50	-14	-24			

Average change for the whole group:
MHI +21 percent and MQLI +2 percent.

COMMENTS

A continued decentralization of the health care sector requires better methods for the evaluation of its activities and results. This applies to methods for finding relevant descriptions of achievement and quality as well as to methods for comparisons and compilations of gathered data. In this paper we illustrate and exemplify methods available to solve this type of problems. These measurements are a complement to other types of information in the decision making process in health care. The result of a change may differ depending on whether quality measurements are included or not. A comparison may also prove to be to the advantage of other entities, individuals or treatments depending on whether quality measurements are included or not.

Regarding HQL, individuals often have different preferences, e.g., some prefer playing golf to collecting stamps, which means that illness and accidents and medical measures affect individual preferences differently. The effect of a fractured left arm is more serious for a golfer than for a stamp collector. It is vital that the method used consider this and does not assume that certain features or activities have a fixed and uniform value. The method presented in this paper does not make such an assumption.

BIBLIOGRAPHY

[1]   Bergner, M.: 1989, 'Quality of Life, Health Status and Clinical Research', *Medical Care*, 3, 148-156.

[2]   Björk, S.: 1989, 'A strategy for assessing quality of life', in Björk, S. and Vang, J. (eds.) *Assessing quality of life*, Linköping Collaborating Centre for WHO Programmes, Linköping, Sweden.

[3]   Albinsson, G., Arnesson, K., Friberg, H. and Olsson K-E.: 1992, *Effects and Costs of total Hip-replacement in Karlskrona and Karlshamn (Effekter och kostnader av total höftledsplastik i Karlskrona och Karlshamn)*. FoU-rapport. HSF (The Council of Health Services Research/Rådet för hälso- och sjukvårdsforskning), Lund, Sweden.

[4]   Jachuck, S. J., Brierly, H., Jachuck, S., and Willcox, P. M.: 1982, 'The effect of hypotensive drugs on the quality of life'. *Journal of the Royal College of General Practitioners*, 32, 103-105.

[5]   Nordenfelt, L.: 1994, 'Towards a Theory of Happiness: A Subjectivist Notion of Quality of Life', in this volume, pp. 35-57.

[6]   Patrick, D. L. & Erickson P.: 1992, *Health Status and Health Policy: Quality of Life in Health Care Evaluation and Resource Allocation*, Oxford University Press, New York.

[7]   Read, J.L., Quinn, R.J., Berwick, D.M., Fineberg, H.V., Weinstein, M.C.: 1984, 'Preferences for Health Outcomes, Comparison of Assessment Methods', *Medical Decision Making, 3*, 316-329.

[8]   Roos, P. and Björk, S.: 1992, 'Health State Index and Quality of Life Index — A Non-Parametric Approach', IHE, Working Paper 1992:7, Lund, Sweden.

[9]   Sen, A.: 1985, *Commodities and Capabilities: Lectures in Economics: Theory, Institutions, Policy*, North Holland, Amsterdam.

[10]   The EuroQol Group: 1990, 'EuroQol — a new facility for the measurement of health-related quality of life', *Health Policy, 16*, 199-208.

[11]   Torrance, G. W.: 1986, 'Measurements of Health State Utilities For Economic Appraisal', *Journal of Health Economics, 5*, 1-30.

[12]   Ware, J. & Sherbourne, C. D.: 1992, 'The MOS 36-Item Short-Form Health Survey (SF-36)', *Medical Care, 6*, 473-83.

ANTON AGGERNAES

ON GENERAL AND NEED-RELATED QUALITY OF LIFE: A PSYCHOLOGICAL THEORY FOR USE IN MEDICAL REHABILITATION AND PSYCHIATRY

1. WHY HAS THE CONCEPT 'QUALITY OF LIFE' BECOME IMPORTANT IN ADMINISTRATIVE AND CLINICAL HEALTH WORK?

Many kinds of treatment are so stressful or risky that the patients ask whether the treatment or cure is worse than the disease. What these patients want to know is what is better or worse for the quality of their lives, not just for their health. Potent forms of therapy with severe side-effects have made it obvious that for most people it is more important how their *lives* are than how their *diseases* are.

During the 20th century, the pattern of diseases has changed. The majority of health work now concerns diseases which we cannot cure completely. This means that two other goals of treatment come to the fore: to alleviate pain and suffering, and to help patients to achieve a better quality of life via rehabilitation, better housing, orthopedic services, etc.

The organization of health services is important. Is it better with public than private health insurance? Large or small hospitals? Community psychiatry or hospital psychiatry? For patients, tax-payers and politicians it has now become more clear that the decisive factor here ought to be what produces most quality of life for the money. This does not, of course, exclude other ethical considerations, for instance, concerning equality and equity.

Many kinds of modern treatment are so expensive that resources are too restricted to provide optimal treatment for everyone. Again, it is important to ask: How much improvement of quality of life can be obtained, for which patients, given the treatments in question?

2. DIFFERENT CONCEPTS OF QUALITY OF LIFE ARE POSSIBLE AND LEGITIMATE

Nobody has a monopoly to define the term 'quality of life' (QoL). Different normative choices are possible. You can claim that QoL is the degree to which you live in peace with your God. Or that the optimal QoL is a state without needs (Nirvana). From some ethical or existentialist points of view, one may claim that the life of every human being is of equal value. As a

L. Nordenfelt (ed.), Concepts and Measurement of Quality of Life in Health Care, 241–255.
© 1994 *Kluwer Academic Publishers. Printed in the Netherlands.*

consequence one may find it unethical or without meaning to talk about degrees of QoL. The quality of lives may also be measured in terms of how often people perform acts which are good for other people, with possible corrections for harm done to others.

In Western culture the most common use of the term 'QoL' is found in discussions of *the degree to which a person feels satisfied and/or happy in life*. Such a definition is, of course, also normative, at least in the sense that it presumes our culture's great emphasis on individual happiness and life satisfaction.

To use or accept this definition does not, however, imply that nothing else in a person's life (other than his QoL) is of value.

3. THE CARDINAL CONCEPT: GENERAL SUBJECTIVE QUALITY OF LIFE (GSQOL)

In my book on quality of life [1] I have defined GSQoL as the degree to which a person feels happy and satisfied in life. Bradburn [5] and Campbell *et al.* [7] also include both happiness and satisfaction in their concepts of general QoL. Nordenfelt [17] only includes happiness. Campbell *et al.* [7] know that happiness and life satisfaction do not always correlate completely in empirical research. There is a tendency for young people to be relatively a bit more happy and a bit less satisfied than old people in sample surveys. This seems to be a special case of the more general fact that level of aspiration can vary — and usually is corrected downwards in old age. In general, however, the correlation between happiness and life satisfaction is high.

Ideally, a person's QoL should be recorded by asking him one or two questions concerning his general happiness and life satisfaction, say on a scale with 3, 5 or 10 steps. In practice, however, it has not been possible to obtain reliable results with such a simple technique. People change their level of aspiration from time to time. Even the basis of comparison may change from person to person and from time to time. Campbell *et al.* [7] mention six possiblities here. It can be a question of 1) the best you could want, 2) the best you could expect, or 3) the best you have tried before; or the basis of comparison may be 4) how people live "in general", 5) how one's relatives live, or 6) how one's closest friends live. Furthermore, important random variation can be ascribed to the fact that a subject on one occasion answers from considering one aspect of his life — and on another occasion from considering some other aspects of his life. Finally, when people evaluate their general QoL they may differ (from others, and from themselves before) as to the relative emphasis on, a) having a good life to look back on, b) having a good life right now, and c) having a good life to look forward to !

The difficulty in recording GSQoL in a reliable and clearly defined meaningful way has made it tempting to divide the GSQoL into components. (In section VIII we return to the problem of giving relative weights to the components. Here it shall only be stated that an ideal solution does not exist. Nevertheless, for some specified purposes it is more reasonable to use questionnaire measurements than to rely on vague estimates.) What really seems to be problematic is to make generalizations from results obtained with questionnaires not firmly rooted in a psychological theory on quality of life. In sections V and VI, I present such a psychological theory. But first a modern concept of health should be introduced.

4. HEALTH, SEEN AS BEING A RESOURCE FOR QUALITY OF LIFE

WHO's definition of *health* as physical, psychological and social well-being is nearly identical with general subjective *quality of life* as defined in section III above. In 1984, WHO [23] had changed from seeing health as quality of life, to seeing health as being some (or all) resources for quality of life. Brogren [6] considers health as social, personal and physical resources "to realize aspirations and satisfy needs". This means that all resources count. High intelligence is more healthy than average intelligence. Average self-confidence is less healthy than strong realistic self-confidence. Anderson [3] finds the concept of positive health important. (Below, I present examples from clinical work of the importance of recording positive health factors in our case sheets.) In WHO's paper [24] concerning "regional targets" environmental health factors are included. In my book on quality of life [1] I have taken the full consequence by defining *health* as being (the degree of) *all kinds of resources in a person and in his environment* for a good quality of life.

This way of thinking was originally inspired by my social psychiatric work with chronic schizophrenics. There were five components in this work:

a) to reduce pathology,

b) to help the patients to a better use of positive health factors in themselves — intelligence, special talents, remaining social skills, etc.

c) to provide new or better resources in the patients — new interests, hobbies, education, social skills, etc.

d) to help the patients to use their environments better, say by telling the patients about job possibilities, housing and day-center facilities,

e) to provide new environmental factors, say teaching of the patient's family or friends how to help him and interact with him, creating patient clubs, sheltered workshops, etc.

In the work with chronically ill or disabled persons there is very often much more quality of life to be achieved via b, c, d and e, than via a. Therefore the very broad definition of the concept of health makes sense. It makes it an obligation for the health team not only to search for disease and health defects in the patients, but also to record the positive health factors in patients and in their environments which can be used in rehabilitation to the end: better quality of life for the patients.

With the presented broad conception of health, it goes without saying that medical doctors are only one among many professional groups with relevant knowledge for health work.

The presented way of thinking is not only relevant to psychiatry. For instance, the Danish Institute for Social Research has done extensive research concering the lives of physically disabled people [2] [12]. The probability of successful vocational rehabilitation was to some extent dependent on the degree of the disability. But at most degrees of disability the following factors seemed equally important: general intelligence, extent and broadness of education, and degree of general self-confidence. The size and the quality of social networks are, of course, also important factors in both vocational and other kinds of rehabilitation.

5. FROM HEALTH TO QUALITY OF LIFE

How does good health contribute to, or lead to, a good quality of life? To live one's life is a process. Health factors, therefore, only contribute to a good quality of life if they are used in the processes of living. This, again, is a matter of satisfying one's needs, especially the fundamental human needs (see below). In Nordenfelt's terminology [16] it is a matter of living up to one's goals in life, especially the vital goals.

In work with chronic patients I have found two reasons to use the concept of needs:

1) Many patients, psychiatric and others, have very vague or unspecified ideas about what they want in their lives. These patients, and others, expect the physician to be an expert in psychology and quality of life. They want the doctor's help in formulating the goals of treatment. The doctor who has a clear conception about the fundamental human needs of all people can communicate this conception to the patient directly. In this way the patient

gets a chance to accept, reject or supplement the value system inherent in the list of fundamental needs. In doctors' advice to patients, their own value systems will always be inherent. To make the value systems explicit minimizes the risk of manipulation and illegitimate guardianship.

That psychiatrists and their patients often have different opinions on which goals of therapy are most important has been shown by Dimsdale and others [8] [9]. Ethically the doctor ought, of course, to respect it if a patient has got values, needs or vital goals which are not included in the doctor's standard conceptions.

Some goals due to paranoid conceptions may be exceptions.

2) A clear and specified list of fundamental human needs is of heuristic value when you, together with your chronic patient, search for ways to a better quality of life. Since Duncker's findings [10] we know that it is vital in problem-solving to specify the goals to be attained. And needs are characterized by their golas (see later). The objects, however, which can be used instrumentally to reach the goals of the needs are within wide limits interchangeable with other objects.

An example to illustrate te point:

A speedway driver may find it extremely important to be one of the best drivers in the world. Now, suddenly his vision becomes defective in a way which makes elite speedway driving impossible. This results in a severe sorrow or depression. If we see this man's speedway driving as a vital goal in its own right, and not as just being instrumental to more vital or basic needs, we will not be able to help him. But if we realize that speedway driving can be seen as just one possible object, among others, then the need for driving is not in itself vital or fundamental, but subsidiary to (Murray [15]) or instrumental to more fundamental human needs, which may be satisfied in other ways. If such other ways are to be found by means of rational thinking, the fundamental needs and their goals should first be specified.

My clinical work has led me to the following list of fundamental human needs [1]:

A)   The elementary biological needs.
B)   A need for warm interaction with other human beings.
C)   A need to have some meaningful activities, felt to have value beyond oneself.
D)   A need for a varied and to some extent exciting and interesting life — essentially a need not to be bored.

Returning to the example:

In one speedway driver the needs A, B, and C may be satisfied in other ways, so that the defective vision only frustrates his need D. For another, the driving gave him nearly all his human interaction. For him, the decisive frustration is of the need B. For a third driver it may be his complete basis of existence (inclusive of A) which is threatened. The visual defect may be the same in these three drivers. But they should be helped in very different ways. The analysis which results in pointing out the frustrated needs is the first step on the way to finding substitute ways to satisfaction of the needs.

Especially in the case of psychiatric patients it is common that they cannot specify or understand why they are sad or depressed. Then the doctor's knowledge of fundamental human needs can be used as a search model to decide which needs are frustrated.

A common problem in the case of many young schizophrenics and many old single women is that they (for different reasons) do not realize how important the need for warm human interaction is for them. If you succeed in having them engage in some kind of warm interaction in curcumstances which do not frighten them, then they may be helped to realize their need. After this first important step the patient is ready to find or create real-life situations in which to satisfy the need in the future.

Young patients with brain damage following traffic accidents may often regain their intellectual and other abilities. But many of their formerly used ways or methods to live a good life may be destroyed forever. One patient has lost his ability to play chess. Another has interrupted his academic carrier. A third one may be forced to leave the labor market completely. In the case of the majority of such patients it is possible to help them to find other activities which may give them a satisfactory life after all. But they clearly need help to formulate their needs and to find objects or ways leading to future need satisfaction. In the case of this patient group it is especially evident that the doctor should know the needs which are common to every human being.

6. A PSYCHOLOGICAL MODEL: CONNECTING THE CONCEPTS OF NEEDS AND EMOTIONAL/HEDONIC QUALITIES

Is happiness an emotion and satisfaction a cognition? This question is based on the classical delusion that cognition, emotion, and volition are separate psychological phenomena. In Fridja's [11] modern conception emotions are experiences with both hedonic and more or less precise cognitive components or qualities. Already in 1914, Shand [20] realized the conative aspects of most emotions.

To the hedonic/emotional sphere belong *emotions, feelings, moods,* many kinds of *sensations* (pains, bodily pleasures) and *sentiments.* The borderlines between these five kinds of hedonic phenomena are not clear-cut. Many actual experiences are emotions and feelings at the same time.

Emotions (see [19]) are *hedonic qualities or experiences which are manifestations of needs being in specified states.* Your emotions inform you of the states of your needs. Prototypical examples are *fear* or *anxiety* in relation to threatening frustration of a need, *sadness* or *despair* in the state of actual frustration of a (fundamental) need, and *joy* or *relief* in the state of statisfaction. When a need is gaining strength (hunger, thirst, sex) it is usually connected with positive hedonic qualities — namely in most cases when you feel confident that the need can be satisfied in the not-too-distant future.

Feelings are *hedonic qualities or experiences in relation to objects* which have, have had, or may have positive or negative value for one or more of one's needs. Fridja [11] uses the word 'concerns' in the sense in which I use the word 'needs'.

Moods are more long lasting hedonic states which are rather vague in their reference to particular needs or objects.

Hedonic *sensations* usually have a rather precise reference to the needs (say sex need) or objects (say pain if the biological avoidance needs are frustrated).

Sentiments (see [20]) are systems of feelings in relation to essential objects, (your wife, your job, a good friend or your country or neighborhood); or in relation to an enemy or rival. Love for one's wife can manifest itself in feelings of *anger* toward people who threaten her, *pleasure* in making her happy, *fear* of losing her, and *sorrow* if she wants a divorce.

In principle, all the mentioned kinds of hedonic qualities can be classified in the following way:

1) *As to which* **phase** *the needs are in;* (see examples under the head "emotions" above.)

2) *As to which* **need** the emotions are related to; say need for food, for human contact, for variation and excitement.

3) *As to which* **objects** are involved. Some kinds of objects are mentioned under the head "sentiments" above. Others are one's own special talents, and one's health.

Finally, it should be mentioned that the hedonic qualities may point exclusively, or more or less, in the direction of the past, the present or the future.

(My conception of need is inspired by the Danish psychologist K.B. Madsen and by works [13] and [19].)

By the term '*need*' I mean *a psychological structure* (biologically based in the brain) *which from time to time gives power and direction to experiences and behavior which are goal-directed in a specified way.*

Different needs are fundamentally different in having different goals or ends. But one and the same need can usually be satisfied in different ways or through use of different objects. *Objects are interchangeable with each other.* Most needs, especially "needs for objects" are only instrumental to or subsidiary to [15] other needs which are more fundamental. The most *fundamental human needs* are not interchangeable with each other, as they are defined by a) being present in all human beings, b) that it for each of them is so that the person in question suffers if the need gets no satisfaction at all, and c) that at least sometimes the need is activated without being subsidiary to or instrumental to other needs.

In childhood and adolescence we learn that only some (among a lot of theoretically possible) objects for each fundamental need are suitable and socially acceptable. In most cultures it is also taught that one ought to be rather stable in most object relations, say marriage, job, friendships. This makes the objects prominent in the feelings in everyday life. For quality of life it means that many people find it to be primarily a question of possession of goods. Only when changes in one's health or in one's environment force one to change objects does it become evident that the fundamental needs are more fundamental than the needs for objects or goods: It is often very difficult to change objects, but it is usually possible.

In my list of fundamental human needs A-D in section V, I did not include a need for security. I find that people feel safe and secure if the needs A through D seem to be reasonably satisfied for themselves and their loved-ones. In many ways you can secure future need satisfaction via possession of objects. In this way the need for security is subsidiary to other needs.

Among the elementary biological needs (A) are the needs of harm-avoid-ance and nox-avoidance with regard to one's own body. The goal situation for such needs is the static one: not to be hurt or harmed. *But for all the other fundamental needs in A through D it is characteristic that their goal situations are processes.* Happiness is not to possess food, but to eat it; not to have friends and a sweetheart, but to interact with them; not to have a job, but to perform it; not to have a hobby, but to perform it so it can yield variation and excitement in life.

The inherent basic philosophy is that the meaning of life is to live it. The presented psychology may be criticized in different ways.

Nordenfelt [16] points out the risk that the list of fundamental needs can be interpreted in too rigid a way. This certainly is a risk — but it need not be inherent. Research may refine the list. For practical purposes the strength of the presented model is that it makes possible a constructive use of pa-tients' feelings and emotions in guiding social medical work.

There are interesting parallels between the presented concepts and Nordenfelt's. We both see the hedonic/emotional area as cardinal in quality of life. Nordenfelt's concept of wishes includes wishes which are not fully conscious at the moment. In this way it resembles my concept of needs. The goals embedded in my fundamental needs are similar to Nordenfelt's vital goals.

7. MEASUREMENT OF QUALITY OF LIFE

Even if only a few examples have been given, it should by now be obvious that the presented concepts of quality of life, health, emotions, feelings, needs, and objects can be useful in clinical work. Notice that this is the case even in some patients who may have fundamental needs which are not, di-rectly or indirectly, included in my list A through D. In routine clinical work it is not, however, important whether quality of life can be measured.

For some clinical conditions different treatments with different effects, side-effects, and risks of complications are possible. For the guidance of a new patient in the choice between treatments X and Y, it would be a value to know whether X or Y generally, or on average, has resulted in an increase in quality of life in former controlled clinical trials. In such clinical trials it is important to measure. Quantification is also important in guidance con-cerning the organization of health services (see section I). For many hu-manists it seems ridiculous to measure quality of life. But when humanists fall ill, they too seek guidance based on knowledge about what has been best for former patients with similar disease patterns. And such guidance presupposes that former patients have been asked about the effects of treatments on their quality of life. Most people are able to tell whether their life in general has become more or less better or worse, or maybe remained unchanged. To indicate degrees here *is* to measure.

8. WHICH KINDS OF INSTRUMENTS SHOULD BE USED ?

In sections II and III the importance of measuring general subjective quality of life (GSQoL) was mentioned. Important problems, especiallly concerning changing levels of aspiration, were found concerning such measurements. My conclusion is that for some purposes it still may be important to ask questions concerning general happiness and life satisfaction, but that other questions often can give equally important and more reliable answers.

The reasoning in section VI leads up to the following concept: *emotion-specified quality of life* (ESQoL) defined as *the total balance between positive and negative hedonic qualities* in a certain time period, say "during the last month". To record this ESQoL you have to question subjects about the occurrence, frequency and intensity of emotionally colored experiences such as anxiety, fear, joy, sorrow, general feeling of vitality, or comfort. It is impossible to include all possible kinds of emotions and combinations of them. In my book [1] I describe a rather primitive questionnaire concering ESQoL (erronously called GSQoL there). It contains six questions concering basic emotions as to occurrence and frequency during the last month. During such a month, however, the actual emotions in relation to present needs are not the only hedonic qualities. Therefore I included in the ESQoL questionnaire one question on whether the subject felt that he or she had a more or less good or bad life to look back on. Finally, two questions concerning optimism/pessimism concerning the future were included. With such a simple questionnaire with nine questions, allowing five steps in each answer, it proved possible to obtain reliable results. The results were also meaningful in the sense that they clearly differentiated suffering psychiatric patients from normal controls.

Given the theory that most needs in most persons' life phases are only subsidiary to or instrumental for the four kinds of fundamental human needs noticed above, then one should expect that a measurement of the degrees to which these fundamental needs are satisfied during the last month should correlate with measurements of ESQoL in the same persons. I did indeed find [1] in examinations of 10 normal subjects, and 20 psychiatric patients, a very impressive co-variation between ESQoL results and the results with a developed questionnaire for *need-specified quality of life* (NSQoL). NSQoL was measured with 12 questions, three concerning each of the four fundamental need areas. The test retest reliability of the NSQoL questionnaire was 0.9, and it is reasonably sensitive to record *changes* in quality of life over time.

In section VI it was mentioned that the hedonic experiences might also be classified in accordance with which *objects* one's feelings are related to. Campbell's *et al.* [7] extensive investigations were made in this way. They used 12 to 16 object areas or life domains, among these "family", "friends", "housing", "job", "health". Subjects were asked how satisfied they were with their situation in these areas, and scores were summed up to a *life domain specified quality of life score* (LDSQoL).The results obtained seemed reasonable in the sense that they co-varied rather well with a score for general life satisfaction.

In computing total scores on ESQoL, NSQoL and LDSQoL it is a serious problem that summing up of scores from different questions has to be done. Should all components be given the same weight? If different weights should be given to different components (before summing up), should it then be the same relative weights used in all subjects ?

In theory, the simple solution is to let each subject at each examination indicate the relative importance of each component for his actual well-being. In practice, however, all investigators of quality of life have found it impossible to obtain reliable results in this way. Nordenfelt [16] emphasizes how different people are in what they judge important. It could be added that especially very ill or very disabled persons may evince especially idiosyncratic weighings of quality of life factors. In theory, therefore, the problem remains important, in practice, it may be less important. Both Campbell *et al.* [7] and I [1] have found that within rather wide limits,nearly the same ranking order of examined subjects will result whether weighted or unweighted scores are used.

It seems fair to conclude that if rather homogenous populations are examined with conceptually clear questionnaires before and after interventions in controlled clinical trials, then it is not only acceptable,but of potentially great value, to include quality of life among the variables to be evaluated.

9. HEALTH RELATED QUALITY OF LIFE

Patrick and Erickson [18] and Torrance [21] strongly advocate employing the concepts of "health-related quality of life" (HRQoL) and "utilities". The basic assumption is that every state of disease or disability can be rated as having a given and constant utility value in a person's life. The utility value is not decided by patients, nurses or physicians, but usually by average tax-payers' guesses about how bad it is to have the condition in question.The great problem here is that we know that psychologically it simply is not true that the same disability always means the same degree of reduced quality of life. Recall the Danish investigation of physically disabled people [2] [12]. Many other factors than degree of disability were important for their functioning and well-being, for instance, intelligence, education and self-confidence.

An example can illustrate the point. A severe hip arthritis may on average reduce quality of life from 1.0 to 0.8 on a scale from 1.0 to 0.0. A given hip operation can, on average, increase the quality of life to 0.9. Should such changes be used to decide whether the hip operation in question should be found in the therapeutic repertoire of surgical departments? Ethically the answer is, of course, no; for there might be a subpopulation of patients who, via the operation, could have their quality of life increased from 0.4 to 0.9. Different health resources, just like all other types of resources, simply are of very different value in the lives of different persons.

Nevertheless, a rather rigid thinking in utility concepts is basic to such popular quality of life scales as the Sickness Impact Profile (SIP), the Quality of Well-being Scale (QWB), and the scale used by Williams [25] to measure quality adjusted life years (QALYs).

In critical discussions of scales for quality of life (see, for example, [4]) proponents (mostly economists) of the concepts of HRQoL and QALYs seem to realize the rather rough nature of the thinking. Their defense runs along the following lines: In times of economic crisis you have to choose among different kinds of treatment, and then rough measures are better than no measures. I do not object to this, but it does not mean that you have to base your reasoning on psychologically inadequate theses.

My own proposal is to measure health factors (resources) separately and quality of life, clearly defined, separately. Then it becomes possible to measure whether a given increase in physical health actually results in a better quality of life; or, if it would have been better to use the money in rehabilitation work, education, or better housing, given that they can be shown to increase quality of life.

10. THE USE OF MEASUREMENTS OF QUALITY OF LIFE IN THE PLANNING OF HEALTH CARE: AN EXAMPLE

In 1990, a special community psychiatry service was offered in one part of Copenhagen municipality. In the rest of the town a more traditional service continued. Evaluation of the services took place in 1992 and 1993; it will be decided whether the special service shall be extended to the whole town, and whether this may be done in a more effective way.

When hospital psychiatry has been (more or less) replaced by community psychiatry, it has often been found to be disadvantageous to the most severely, chronically ill patients. Therefore, in Copenhagen we started the following investigation: Before the start of the community service we investigated all known severely ill, chronic psychiatric patients belonging to the experimental district in respect of a number of disease factors, health factors, and other resource factors besides measuring quality of life with Dupuy's scale for General Well-Being (GWB). (See [14] and my scale for need specified quality of life (NSQoL)). In a control district, the same investigations were made concerning the known severely ill patients belonging to that district. In 1992 and 1993, all examined patients were to be reexamined in respect of the same variables. If the GWB scores, on average, in the experimental district remain unchanged (or have increased or decreased in the same way as in the control district), it would indicate that the chronic patients had not received a poorer service in the community psychiatry system. This conclusion would be strengthened if the NSQoL scores were unchanged, too.

That we evaluated disease factors, health factors, and quality of life scores independently made it possible to generate hypotheses about which changes in health may have caused which changes in quality of life. That the NSQoL scores have a separate value, besides being a check on the validity of the GWB scores, can be illustrated by the following example. Say that we find a moderate increase in both GWB scores and NSQoL scores after the three years of community psychiatry. Then one would want to know whether one could do even better in these new centers. Maybe the four NSQoL components, corresponding to the four kinds of fundamental needs, have changed differently, so that the patients had had poorer satisfaction of the needs for warm human interaction and for meaningful activities. However, this was reflected in the sum-scores and was more than compensated for by an increase in satisfaction of the elementary biological needs and the need for variation and excitement in one's life. Then one would, of course, in future centers, try to provide more facilities in the need areas where decreased quality of life had been found.

CONCLUDING WORD

Rational planning presupposes that one use concise and separate concepts concerning various kinds of quality of life, on the one hand, and concise and specified concepts of illness factors, positive health factors, and other relevant resources on the other. Finally, treatment goals concerning chronic patients have to be specified in terms of their quality of life when one cannot cure their disease.

BIBLIOGRAPHY

[1]   Aggernaes, A.: 1989, *Livskvalitet: En bog om livskvalitet som centralt begreb i sundhedsarbejde, socialt arbejde, kulturdebat og politik*, FADL's Forlag, Copenhagen.

[2]   Andersen, B.R.: 1966, *Fysisk handicappede i Danmark. IV: Arbejde og erhverv*, Socialforskningsinstituttets publikation 22, Teknisk Forlag, Copenhagen.

[3]   Anderson, R.: 1984, *Gesundheitsförderung: Ein Überblick*. In Baric,L. (ed.): Europäische Monografien zur Forschung in Gesundheitserziehung, *6*, Bundesministerium für Gesundheiet und Umweltschultz, Vienna.

[4]   Baldwin,S., Godfrey, C., and Propper, C. (eds.): 1990, *Quality of Life. Perspectives and Policies*, Routledge, London and New York.

[5]   Bradburn, N.: 1969, *The Structure of Psychological Well-Being*, Aldine Publishing Company, Chicago.

[6]   Brogren, P.-O.: 1985, *Promotion of Mental Health*, Rapport NHV 1985:3, Social-departementet, Stockholm.

[7]   Campbell, A., Converse, P.F., and Rodgers, W.L.: 1976, *The Quality of American Life. Perceptions, Evaluations, and Satisfactions*, Russell Sage Foundation, New York.

[8]   Dimsdale, J.E.: 1975, 'Goals of Therapy on Psychiatric In-patient Units', *Social Psychiatry*, *10*, 1-7.

[9]   Dimsdale, J.E., Shershow, J.C., Klerman, G.L., and Kennedy, A.M.: 1978, 'Social Press and its Influence on Psychiatric Treatment Goals', *Social Psychiatry*, *13*, 153-157.

[10]   Duncker, K.: 1935, *Zur Psychologie des produktiven Denkens*, Springer, Berlin. English version:'On Problem Solving', *Psychological Monographs*, *58*, no. 5, 1945.

[11]   Fridja, N.H.: 1986, *The Emotions*, Cambridge University Press, Cambridge.

[12]   Kühl, P.H.: 1967, *Fysisk handicappede i Danmark VI: Psykologiske forhold*, Socialforskningsinstituttets publikation 27, Teknisk Forlag, Copenhagen.

[13]   Maslow, A.H.: 1968, *Toward a Psychology of Being*, 2nd ed., D.van Nostrand Company, New York.

[14]   McDowell, I. and Newell, C.: 1987, *Measuring Health: A Guide to Rating Scales and Questionnaires*. Oxford University Press, New York.

[15]   Murray, H.A.: 1962, *Explorations in Personality* (1938). Science Editions Inc., New York.

[16]   Nordenfelt, L.: 1991, *Livskvalitet och Hälsa: Teori och Kritik*, Almqvist & Wiksell, Stockholm.

[17]   Nordenfelt, L.: 1987, *On the Nature of Health: An Action Theoretical Approach*, Philosophy and Medicine 26, D. Reidel Publishing Company, Dordrecht.

[18]   Patrick, D.L., and Erickson, P.: 1988, 'What Constitutes Quality of Life ? Concepts and Dimensions', *Clinical Nutrition*, 7, 53-63.

[19]   Rasmussen, E.T.: 1960, *Dynamisk psykologi og dens grundlag*, University of Copenhagen, Copenhagen.

[20]   Shand, A.F.: 1962, *The Foundations of Character* (1914), Akademisk Forlag, Copenhagen.

[21]   Torrance, G.W.: 1986, 'Measurements of Health State Utilities for Economic Appraisal. A Review', *Journal of Health Economics*, 5, 1-30.

[22]   WHO: 1948, *Constitution of the WHO*. In: Basic Documents, 15, WHO, Geneva.

[23]   WHO: 1984, *Health Promotion. A Discussion Document on the Concept and Principles*, WHO, Copenhagen.

[24]   WHO: 1984, *Regional Targets in Support of the Regional Strategy for Health for All*, WHO, Regional Office for Europe, Copenhagen.

[25]   Williams, A.: 1985, 'Economics of Coronary Artery Bypass Grafting', *British Medical Journal*, 291, 326-329.

MADIS KAJANDI

A PSYCHIATRIC AND INTERACTIONAL PERSPECTIVE
ON QUALITY OF LIFE

INTRODUCTION

A modern view in psychology stresses the interaction between individual and environmental factors as an important basis for analyzing psychological events. This view, which has been called interactionism, has developed in complexity over the years, and now a physiological level of variables is often also included in the analysis. The interactional perspective has spread into many fields of application, and in clinical psychology Magnusson and Öhman [18] discuss the possibility of applying it to phenomena connected with psychopathology.

The interactional perspective has also expanded into the field of research on quality of life. For example, Murrell and Norris [20] working from a community psychology viewpoint, find it possible to understand the concept of quality of life from two different perspectives. According to a psychological perspective it is a mental health variable indicating the 'goodness in life', and alternately, according to a social perspective, it is an indicator of 'goodness in the society'. By means of the latter it is possible to evaluate to what degree it is possible to achieve the goodness in life. Moreover, in accordance with Bubholz et al. [7] they identify as a possible level of analysis the interaction between individual human needs and the resources of the environment to satisfy those needs. This position, thus, makes an attempt to integrate the human level with an environmental level into an interactional perspective.

In the present article I will outline a research project that has explored the possibilities of measuring the quality of life in psychiatric populations. Efforts to maintain an interactional perspective have been made in the definition of the concept, in the procedures for the measurement of the concept, and in the therapeutic implications that can be inferred from the measurements.

L. Nordenfelt (ed.), Concepts and Measurement of Quality of Life in Health Care, 257–276.
© 1994 *Kluwer Academic Publishers. Printed in the Netherlands.*

1. THE UNIVERSE OF THE GOOD LIFE

Research on quality of life has increased dramatically in the last decade. In a survey of literature, Diener [10] examines over five-hundred articles on this topic. The dramatic increase probably has much to do with the need for valid criteria variables in both research and clinical work. In human activities concerned with providing care for people, the concept of cure is a natural variable for deciding when one has succeeded and when one has not. But the need for other kinds of criteria increases when complete cure cannot be obtained. In such cases the care has to provide possibilities for a life that is as good as possible although some functional deficiencies remain, and this, in turn, results in a need for matching criteria.

Off course this is nothing new. Care-giving activities have always been directed to optimizing the good for people. But the "new" seems to be that 'the good life' in itself is investigated more directly and explicitly. If this concept in terms of quality of life can be given a specified and suitable definition, it can provide guidelines for care-givers that are more exact and more in agreement with one another than is presently the case. Not least important, more explicit definitions can be scrutinized and discussed between clinicians, researchers, and patients. There are also economic reasons for developing proper criteria. In times when resources are restricted, the inherent conflict between authorities that allocate resources and care-givers and patients who use these resources tends to increase. Therefore, reliable criteria for what is good care and what is not can help us make better decisions about which kinds of care activities to support.

However it is not an easy task. Although research about the concept of quality of life has been operating for many years, and although much work has been done, this concept still resists a precise definition. In this stubbornness also lies much of its fascination.

To bring some order into this complex matter let us for a moment imagine a universe that includes all possible indicators of 'the good life'. Until today we have identified some indicators, but certainly many still remain to be detected. The most well-known concepts in this universe are perhaps the concepts of *happiness* and *health*. These two were identified already in ancient times but also remain important today. Nowadays, and in more empirically directed research, the happiness concept often is split up into a social aspect, *welfare,* and a psychological aspect, *well-being,* although it still earns a place of its own in the universe.

The three concepts — welfare, well-being, and health — also represent different scientific areas. The welfare concept is often connected with sociological research, where it is used to describe living conditions in societies. The well-being concept is often connected with psychology and is used to capture different pleasant cognitive and emotional states. The health concept is connected with the medical research area and is used to describe the differences between illness and health and the functional losses caused by illness (see, for example, Bergner *et al.* [4]). Sometimes, however, the health concept is defined in terms of possibilities of fulfilling personal goals (see Pörn [23]), and in such a case the health concept also has a psychological significance.

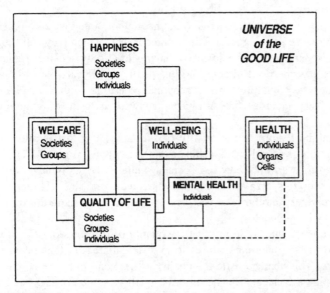

FIGURE I. Interrelations between concepts in the universe of the Good Life

In the beginning of the present project we examined the extant literature regarding the concept of 'quality of life' [15]. We found that it had close connections to existing concepts in the universe of the good life. In fig I an effort is made to illustrate our interpretation of how different authors have defined and used the quality of life concept in relation to other concepts. Many examples can be found of how authors use the term 'quality of life' as a welfare concept, and the same is true of the well-being concept. Also, the health concept is often defined in quality-of-life terms on special scales that measure various impacts of illness. The inclusion of the mental health concept in this context is based on results from an extensive review of the literature by Jahoda [13], where she pointed out the connection between criteria of mental health and psychiatric patients' quality of life.

The quality of life concept thus can be associated with almost all traditional indicators in the universe of the good life. Although defined differently in different sciences there seems to be some common features. When researchers use the term 'quality of life' their objective appears to be to describe and measure something more sophisticated than what had been possible with their former conceptual apparatus. The new concept does not replace the old ones, but supplies something new, beyond what has formerly been measured.

Another characteristic seems to be that authors integrate different aspects of the universe of the good life into one concept, which they call 'quality of life'. A sociologist, while formerly having used only social variables, may integrate aspects of well-being into his quality of life variable and a medical researcher may complete his conceptual repertoire with a quality of life variable that includes health aspects as well as well-being and welfare aspects.

A unique feature of the quality of life concept is that it often includes people's interpersonal relationships as an area of investigation. This feature thus identifies an area between societal welfare and individual well-being. Elements in this area often include the quality of the relationships to a marital partner, family, relatives, and friends, but also more distant relationships may be included.

To sum up, one could say that the quality of life concept often is used to bridge over and integrate features from different areas in the universe of the good life. This characteristic the quality of life concept has in common with the happiness concept. But because the quality of life concept has emerged within the framework of empirical science it has the requirement of measurability attached to it. Thus, it can be said that the quality of life concept and the happiness concept are both similar and different from each other. They share the integrating characteristic but they differ in terms of the empirical bond attached to them. Maybe it can be stated that the quality of life concept is a modern and more empirically directed equivalent to the happiness concept.

But what is really meant by the term 'quality of life'? A natural way to try to provide an answer to this question is given by an analysis of the term in itself. The problem seems to fall into two parts. The first has to do with the specification of which aspects of life are especially important, the *life*-part, and the second has to do with decisions concerning whether or not these aspects of life are good, or, in other words, when they have *quality*.

2. STRATEGIES FOR THE DEFINITION OF 'LIFE'

Sometimes it is proposed that the important things in life are so unique to every individual person that they cannot be given a general definition. The argument is that what is important is so personal that every general definition will produce a bad fit with an individual's private view. Although it may be a fact that most people have a personal view about what constitutes life quality, and that in each case this view is felt by the individual to be unique, it is still possible to argue for the suitability of more general definitions.

Above all it seems reasonable to assert that most individuals, although they *feel* their definition is unique, in fact share an underlying evaluation of what is important in life. Fundamental needs, for example, belong to that category. Some empirical research support this assertion. When experiments are done to determine what respondents include in the concept of quality of life, one can often find that people make similar proposals, which supports the idea of an existing conformity.

Another important argument for general definitions is a practical one. Although personal definitions of life are appealing they present problems in scientific contexts. The problems primarily concern questions regarding possibilities of comparison. It is simply not possible to compare groups of people in a meaningful way if you don't know to what they refer. And if you ask every participant in a study to be precise about his/her preferences in respect of life, one has to make a judgmental ranking of the different proposals afterwards to be able to compare meaningfully. General definitions avoid much of these problems, and provide the possibility for comparison between different groups of respondents and different periods of time.

When specifying life domains one has to decide on a method by which to do this. In the literature one can identify at least three distinct methods. The *personal method* (which we have partly discussed) is probably most common when people reflect upon their lives from day to day, but it is also used in scientific contexts. One way according to which the personal method is applied in scientific contexts is when people view their lives as a whole. This was a ususal method in early quality-of-life investigations [9] [12] [6], but *life as a whole*-estimates, registered on Likert-scales with three, five, seven or ten steps, are still in use in quality-of-life research. The characteristic feature of this type of life-definition is that no overt declaration of what is included in life is made on behalf of the respondent and therefore, in a sense, the problem of specification is avoided. There are aslo examples in the literature where an explicit personal definition of life is asked for from the respondent, but these are rather rare. The problem in common for both the life-as-a-whole-method and the explicit-personal-definition-method; they make meaningful comparisons questionable.

Another method of giving precision to life is *empirical*. According to this method, questions are directed to groups of people regarding what they consider the most important aspects of life. The answers are analyzed and accumulated into increasingly overarching categories. Flanagan [11] asked a sample of normal people to make priorities of life circumstances that influenced their quality of life upwards or downwards. The analyzing process resulted in seventeen life areas and activities. Blau [5], using in essence the same method in a population of patients in psychotherapy, found that this sample gave priority to twelve life domains. When directed to different kinds of subgroups of people, such experiments often result in lists of life priorities that are similar in their core elements although they may differ in some specific aspects.

Difficulties with the empirical method of extracting life priorities have to do with validity. The target population to which questions are directed is seldom given the opportunity to make a thorough analysis of their task and this may result in proposals that are based on immediate responses, which in turn may result in shallow definitions.

Some of these problems can be avoided when a *hypothetical method* is used. The term 'hypothetical' emerges from the fact that in this case the researcher makes a proposal concerning which life aspects are important (based on a particular hypothesis). Previous research, professional considerations, clinical facts, etc. may be different origins of this hypothesis.

Also, hypothetical definitions of 'life' result in lists of life domains but sometimes life values are also included. To include life values makes it possible to transcend the borders of concrete life domains and offers a second dimension of abstract categories. This makes the definition of 'life' less shallow and probably more acceptable to different individuals.

Andrews [3] and Andrews & Withey [2], for example, use life domains in combination with life values. According to these researchers, life is best evaluated by means of estimates in two dimensions and the two dimensions are also used alternately to evaluate each other. For example, when the life domain 'Housing quality' is evaluated, people are asked to consider how much 'Beauty', 'Security' (life values), etc. they can obtain from their present housing. Accordingly, when the life value 'Fun in life' is evaluated, every life domain is supposed to be scanned to find out the degree of fun that the respondent can experience. The sum of all goods in the matrix created by the two dimensions is reflected in a total quality-of-life score.

An important advantage of the hypothetical method is that more funda-mental needs can be included in the definition. Such needs run an obvious risk of being discarded in the personal and empirical definitions because the needs may already be satisfied and therefore forgotten. A fundamental need may also be overlooked because the person in question is unconscious of the existence of such a need, or the importance of it. Two colleagues of mine have pointed out the analogy in this context with the need of vitamins in the body. This need is vital whether we know about it or not, and whether we want it to be so or not. The same may be true of other fundamental needs.

In psychiatry these kinds of considerations are essential. Psychiatric patients often have difficulties in satisfying their fundamental needs by themselves. For example, they may deny the need for social contact, the need for food, or the need for personal hygiene. In part this is caused by their illness, which may overwhelm them to such a degree that they cannot identify the need in question. But for many patients it has also to do with a long-lasting isolation from normal social life. Living in an institution for long periods of time may result in indifference to ordinary social norms.

3. STRATEGIES FOR THE MEASUREMENT OF QUALITY

When one has decided how to define life the task of assessing when the proposed definition includes quality comes into focus. Essentially, this problem involves two decisions. First, we must decide which kinds of rules or criteria are to be followed when measurement is done; then it must be decided who is to apply the rules and perform the actual measurement, i.e., Is it the individual himself or someone else (like an interviewer)?

The rules in question can be *relative* or *absolute*. Relative rules imply that every individual is to decide which norms are to be applied to the concept (subjective measurement), and therefore such rules can vary between individuals, between situations, and over time. Absolute rules, on the other hand, are made up by someone else than the individual (in this context the researcher) and therefore such rules are normative.

When absolute rules are established they are the same rules for every individual, situation and over time, which increases the possibility of undertaking meaningful comparisons. The situation can be compared to a test situation. Because the absolute rules are normative, it is important that the researcher gives access to the guiding principles, interview manuals, etc. so that the norms can be scrutinized and discussed.

Criticism of the measurements based on relative rules has to do with empirical findings of low to moderate correlations between such kinds of measures and more objective descriptions of welfare. This discrepancy, on the one hand, forms an argument in favor of subjective measurement because this shows that there is more information to obtain than is given by the different kinds of objective measures. But on the other hand it also produces a difficulty about which type of measurement one is to base one's decisions on. If the objective indicator points to a low quality of life, but the subjective points to a high, which indicator should be trusted?

Subjective quality measures are highly dependent on specific individual characteristics. Knowledge of alternatives, level of aspiration, and ability to identify one's position in the world are examples of such factors. The Pollyanna phenomenon which means that people in various degrees tend to idealize their life situation, is also included, as is the risk that respondents will lie when they make their judgment on a questionnaire.

Subjective measurement in terms of personal satisfaction has been thoroughly analyzed by Alex Michalos [19]. He identifies eight kinds of comparisons that individuals make to evaluate their life satisfaction. In the center is the discrepancy between what the individual has and what he wants to have. Other important comparisons are made between how life was earlier compared to the present, how other people live compared to oneself, how life is now compared with what is reasonable and fair, etc. Michalos' multiple discrepancy theory (MDT) elegantly and comprehensively offers an in depth analysis of how people evaluate their life satisfaction.

But is life satisfaction really the same as life quality? To a certain extent it is. Obviously, you cannot have a high quality of life and at the same time be dissatisfied with life. The close linking of the satisfaction concept to people's knowledge of alternatives and their aspirations regarding life, however, also results in problems.

Some empirical studies, for example, indicate that people in different, less privileged life situations tend to rate themselves higher on rating scales regarding their subjective satisfaction than do people that are better off in the society. Campbell [8] explains these democratically cumbersome phenomena by saying that rich people and educated people have "reference points for criticism" while their less favored fellow citizens may have acquired "constricted horizons".

Another problem arises as a consequence of life satisfaction being so closely linked to people's aspirations. Especially extreeme aspirations seem cumbersome. Take, for example, an individual that totally lacks aspirations and who consequently does not want anything. Given a life satisfaction scale he should rate himself at the top (if he does not he does indeed recognize something in life that he wants so that alternative is not possible). But can we agree in such result. Probably not? Having aspirations, striving for goals and wanting things in life is something of the essence of human life; a total lack of these things when such a person rates himself having a top quality of life is simply hard to accept. What hampers the life satisfaction approach for quality-of-life measurement is that it cannot handle aspirations that fall outside certain limits. The method seems to presuppose *reasonable* aspirations.

Another consequence of the linkage between life satisfaction and individual aspirations has been pointed out by Næss [21]. If subjective life satisfaction should become the criterion of choice regarding, for example, social policy making, it would be possible to increase people's satisfaction by arguing for modesty with aspirations and by encouraging people to bring up children who demand little from life. Because of problems like these, Allardt [1] argues for the use of combined objective and subjective criteria instead of pure subjective. Decisions based on sole subjective measurements could lead us into a 'hopeless conservativeness', he says.

The problems connected with relative rules are especially apparent among psychiatric patients whose emotional and cognitive abilities may be damaged to a extent that normal aspirations cannot be formed. The result may be both extremely high and extremely low aspirations. Sometimes a patient's aspirations may vary between these extremes in correlation to phases of their illness. Moreover, these people often have spent decades of their lives in mental institutions where they have been cut off from influences and references of normal life. The result is people with "constricted horizons" in the worst sense of this expression.

4. THE PROPOSED DEFINITION OF QUALITY OF LIFE

The discussion above indicates that the use of subjective methods in both defining 'life' and when measuring 'quality' seems to presuppose a certain
degree of maturity, mental health and social integration in the group of
people that is in focus, if they are to appear fully acceptable. Psychiatric patients often lack all of these requirements. Therefore we have arrived at the
conclusion that to make reliable and valid measurements of the quality of
life, we have to find strategies that circumvent the possible bias caused by
patients' mental illness.

Our target population is all kinds of people, healthy as well as different
patient populations. Among the latter, all psychiatric patients form a core
population. The heterogeniosity of this application out-ruled the possibility
of using personal or empirical methods to define 'life'. Instead, using a
hypothetical method, we chose variables from different facets of the universe
of 'the good life' that had both psychiatric and general relevance. By
combining social variables with psychological and interpersonal variables we
maintained the possibility of an interactional perspective.

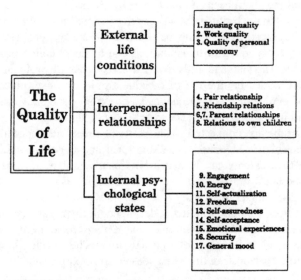

FIGURE II. The structure of the proposed quality of life concept

'Life' is defined as a combination of variables measuring *external life condition, interpersonal relationships*, and *internal psychological states* (see fig II). The *external life conditions* include measurements of 'Housing quality', 'Quality of work life' and 'Personal economy'. These three life domains cover central problem areas for psychiatric patients, and at the same time they are relevant areas for other people.

Interpersonal relationships include relationships to a 'Partner', 'Friends', 'Parents' and 'Own children', while the *Internal psychological states* cover life values and mental health variables like 'Engagement in life', 'Energy', 'Self-actualization' and 'Freedom'. Further are included 'Self-assuredness', 'Self-acceptance', 'Emotional experiences', 'Security' and 'General mood'. The internal set of variables are the same ones proposed by Siri Næss [21]. In addition to these seventeen variables is a variable that measures '*Quality of Life as a whole*'.

When deciding which method to use for quality assessment we again were met with the dilemma of having to choose between subjective and more objective methods. On one hand, there was the obvious risk of biased subjective measurements in our core target population. On the other hand, subjective measurements, when they work, are both democratically attracting and, above all, the sophistication striven for in quality-of-life measurement may be accomplished in a striking way if one succeeds in getting a a reliable subjective measurement.

The solution proposed in this project is to measure the defined concept by means of two parallel methods. One method is purely subjective and the measurement is accomplished by subjects ratings on five-point Likert scales, one for every eighteen variables (including the one for the life as a whole perspective).

In the other method a trained interviewer makes estimates of the respondents quality of life on the same kind of scales and the same variables. In this method, however, the interviewer makes professional estimates that are based on up to three kinds of information, depending of which variable is in focus.

For example, when estimating a patient's 'Housing quality' the interviewer first considers the objective housing condition. Is it an apartment, a group home, or is it a ward at a mental hospital? Second, is considered how well these housing conditions harmonize with the patient's abilities to function, i.e., how well does the patient interact with these particular physical surroundings?

A basic assumption in this context is that a good interaction between the patient and his housing exists when the patient has *maximal autonomy in his housing and at the same time can benefit from a good care quality.* Consequently, one can increase autonomy only to the degree in which good care quality is preserved, or alternatively, it can be said that one can increase the support in the patient's home to uphold a good care quality if the degree of his autonomy at some time should be too high. This inbuilt regulator between degree of autonomy and amounts of care efforts also renders the measurement a dynamic character. A solution that is optimal at one time period may be inadequate at another, so evaluations have to be made on a regular basis.

The third type of information considered when the interviewer makes his estimate, is the patients own subjective evaluation of his situation. According to the rating instructions it is not possible to make a top rating if the patient dislikes the conditions in a particular life area. For example, imagine a patient who lives in a group home that is considered the best solution form the care-givers point of view. If the patient also esteems this solution highly all is good and an interviewer is instructed to make a top rating, but if the patient dislikes the solution proposed by the care personnel the patient's 'Housing quality' is not optimal. Maybe the care personnel has to find the patient another group home or offer another solution that can be accepted by the patient. Above all, people responsible for the care have to approach the patient, create contact, and find out what is the matter with the present housing.

In other life areas, for example social relationships, the three areas take form in other kinds of questions. For example, first is considered whether social relationships exist; second, to what extent the relationships include elements like warmth, closeness, honesty, competence for handling conflicts etc.; third what is the patient's subjective view of his situation regarding social relationships. By combining information from different levels of reality and by considering how these levels harmonize with one another the rating procedure can be said to adhere to an interactional perspective.

The definitions and the rating instructions are written down in a manual, so the interviewer is guided by predetermined normative rules when making his estimates. This is further facilitated by the inclusion of descriptive anchoring-points for the terminal and middle points on the scales.

The normative elements built into the interviewer manual have been formulated by means of an analysis of the concept of mental health [8], the normalization principle [20] [21] and clinical psychiatric praxis. In the formulation of normative principles regarding the variables in the internal psychological states, the rating instructions given by Næss [21] have been helpful, and these have been revised and completed to suit the psychiatric field of work.

The interviewer estimate thus emanates from an assumption of the existence of certain fundamental needs that people have. These fundamental needs are based on value statements and include the need for belonging to a social context and the need for a private area, both physical and psychological. Further, there is the need to have possibilities for formulating personal goals in both long and shorter perspectives and of having the external and internal resources and capacities to actualize such goals. These are all examples of fundamental needs that under various headings are woven into the operational definitions of the variables.

To sum it up; it can be said that the interviewer method of estimating the quality of life requires the satisfaction of both the fundamental needs and the individuals personal needs to reach a maximum estimate in the different variables. The fulfillment of different needs can be examined during the interview by specific questions which is in contrast to the pure subjective questionnaire method. For practical reasons the questionnaire cannot be elaborated with explanations of what different needs are. The result is that fundamental needs are considered only to the degree that the respondent himself identifies such needs. To illustrate the difference let us discuss an example in a little more detail.

Suppose we have a patient with schizophrenia who wishes to live in an apartment of his own. Let us also suppose that he is adequately pharmacologically treated and that his care-givers are prepared to help him to fulfil his wish. In the apartment, however, the patient begins acting out bizarre ideas to the extent that neighbours react. Maybe he also neglects the proper care of himself and the apartment to the extent that people around him, both professional and laymen, start to question the hygienic and medical consequences. When problems accumulate to a certain level responsible care-givers have to take action.

This patient, given the subjective quality-of-life scale, may indicate his 'Housing quality' at the top, and let us for a moment suppose that he makes this rating. Parenthetically, it can be said that according to our empirical experience severely ill patients can indicate themselves in the top values on a self-rating scale in total opposition to a more objective estimate.

Probably the solution to the problem will be to move the patient to a group home or with an increased amount of staff support. But crucial in this context is not the actual solution to the problem but how the solution corresponds to different types of quality-of-life measures, and how an individual's personal view corresponds to important persons' view around him.

In this example the actions taken will clearly be at odds with the patient's subjective estimate, because according to his personal top rating on the quality of life scale no action should be taken. He is satisfied and happy with the housing conditions. But important people around him disagree. This illustrates a limitation that is built into measures that are based exclusively on individual subjectivity. The subjectivity seems to be fully taken account of only if the consequences reasonably correspond to normal social and medical standards. If not, other criteria tend to emerge and the individual's subjective view is depreciated.

An interviewer estimate, according to the method proposed in our project, would not result in a top rating, but probably in a rating somewhere in the middle of the scale. From such a rating it is much easier to argue for a change in the patient's housing situation, and the action taken will not be totally at odds with the quality-of-life measurement. The difference, thus, between the subjective rating and the interviewer's estimate is that the latter corresponds to vital care needs and to societal norms while the former does not.

In a less extreme situation the problem discussed above may have another consequence. Suppose that there is a shortage of group-home places and that politicians are considering building some new group-homes. Met with a subjective quality of life assessment that indicates predominantly top ratings for the existing housing solutions and coupled with financial difficulties, their final decision may be to abandon of the building project.

5. SOME EMPIRICAL RESULTS WITH THEORETICAL SIGNIFICANCE

Inter-rater reliability

In this section I would like to illustrate the discussion above with some empirical findings. The first has to do with the inclusion of a method of quality-of-life measurement that builds on estimates made by an interviewer. This procedure calls for tests of the inter-rater reliability of the interviewer ratings. It is obvious that we must avoid the situation where every interviewer makes ratings according to his own personal view and try to create a situation where interviewers follow the given instructions when they make their ratings.

In our project the method of estimating quality of life by means of an interviewer has been subjected to different controls regarding inter-rater reliability. Both in studies especially designed for that purpose and within the context of other studies the reliability has been quite high. Reliability scores around and above .90 are more the rule than the exception [14]. Similar results have been obtained in other projects.

In an early study, copies of eight audio-recorded interviews were distributed among nine independent raters together with the rating manual. Three raters had a special training in the rating method and were indeed the same persons that had made the interviews; three other were psychology students; the remaining three were laymen. The two latter groups had no special training in the rating techniques but made their ratings solely on the basis of their impressions and the instructions provided in the manual.

The results indicated that intergroup and inter-rater agreements were good, and the results encouraged us to use the instrument in the future. Among factors that counteracted good agreement were interview quality, and the number of interviews and estimations made in succession by the same interviewer. This latter phenomenon, observer drift [16], has to do with the fact that raters, in the long run, tend to develop personal modes of interviewing and rating. The phenomenon is well-known and can be counteracted and handled by means of interspersing repeated retraining sessions with the interviewers within the succession of interviews.

6. COMPARISON OF SUBJECTIVE AND INTERVIEWER ESTIMATES

The subjective instrument has been submitted to three samples of respondents, and the results from these samples can be compared with interviewer estimates made on three comparable groups of respondents. In figures III and IV this comparison can be observed graphically.

The most significant feature of the subjective instrument seems to be that subjects' modes of responding are similar regardless of their background in terms of how stricken they are by psychiatric illness. The overall mean level of subjective quality of life is therefore rated on the same level by the different groups. The only variables that clearly discriminate between groups are the 'Work quality' variable and the 'Personal economy' variable, which are rated significantly lower for a group of psychiatrically healthy respon-

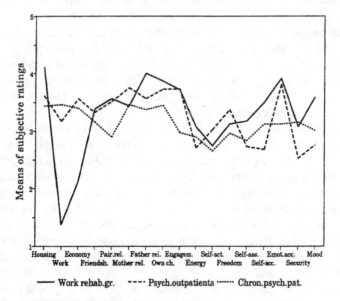

FIGURE III. The subjective quality-of-life ratings of three groups of respondents

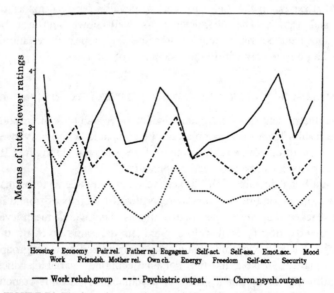

FIGURE IV. The interviewer estimates of the quality of life for three groups of respondents

dents that participated in a work rehabilitation program than for psychiatric groups of respondents.

Other types of background information do have discriminating power in the subjective instrument, but this seems to decrease as a function of a patient's illness. In comparatively healthy groups of patients the subjective instrument does differentiate according to different background information, but in chronic hospitalized groups it does so only to a limited degree. Moreover, the more ill the patient the less chance there is to obtain an adequately completed questionnaire although, as in our project, extensive help was offered by staff.

The interviewer instrument, in turn, does discriminate significantly between groups of individuals with different degrees of psychiatric illness and does so also in respect of other background information. According to our experience it is also easier to approach severely ill people by means of an interview than with a questionnaire. In order to increase such patients' willingness to participate we have made use of the staff in the wards that stood close to the patients. Staff were trained in interviewing and rating techniques and patients were interviewed on several occasions according to their stamina. In a limited number of patients the interview resulted in temporarily increased psychotic symptoms, but since the interviews were accomplished in close connection with the care system these could be easily handled. Especially among units that emphasized patients ability to verbalize emotional material the interview and patient's experience of it could be integrated into the active rehabilitation of the patient.

The results so far indicate that both quality-of-life measures contribute to our understanding of the problems involved but in different ways. The interviewer estimate can be applied to all subgroups of patients and seems to have reliable measurement characteristics, the subjective estimate can be applied to most subgroups of patients and is a direct, fairly easy way to capture alterations in patient's appreciation of life. In designs of evaluation projects the subjective method can be used to follow the process of treatment while the interviewer estimate can be used to evaluate before-and-after changes and differences between groups.

7. CONCLUDING REMARKS

The present approach proposes two different indicators for quality of life. Neither of the two is considered by us to be automatically superior or more truly reflecting reality than the other. However, the interviewer estimate is closely linked to general values which coincide with typical care goals. This instrument will probably be the one that a care-giver with responsibility of offering good care will follow if he or she chooses to use this criteria. However, since the subjective view of the patient also has influence on the estimates resulting from this instrument he cannot ignore what patients want. Thus, in order to offer his patients an optimal quality of life he has to offer care that satisfy both needs seen from a care perspective and the patient's individual needs.

A unique possibility with double quality-of-life measures is offered by equating the difference between the interviewer and the subjective estimates. The more there is a difference the more does the patient's view differ from the views of those around him. Changes in the difference may therefore be used as an indicator in itself.

If there is a difference in views, the first care objective in psychiatry, as I see it, would be to investigate this difference and enter into a sincere human interaction with the patient to find out where and why the difference in view appears. Such human interaction can take place on many levels in the care system, all the way from intensive psychotherapy to day-to-day bedside conversations. In either case the human interaction must start from a fundamental acceptance of the validity of the patients' conception of reality.

If such an interaction between the patient and empathic professionals is successful it may eventually result in a pooled quality-of-life estimate where the patient and his care-giver looks upon things in much the same way. In such cases the care system has succeeded in bridging over the patient's mental isolation or, which is also possible, the patient has succeeded conveying his unique personal conception of life in a trustworthy way to the external world.

Parallel to work on increasing closeness and understanding between the patient and the care-giver, efforts must be directed toward increasing the overall level of the quality-of-life scores and this can be seen as the second objective of the psychiatric care system.

There is an expression, 'the quality of life is in the eye of the beholder', which has been characterizing the major part of quality-of-life research. The expression symbolizes the strong subjective position that has dominated research in this field. The inclination toward individual subjectiveness is not surprising, because the focus of quality-of-life research has to a major part been normal people.

Our quality of life project, which has been presented here, did not have normal people in focus and this led us to another approach. This approach does not presuppose mental health and social integration which the traditional approach does; moreover, it places in focus the individual's interaction with his surrounding. As a consequence, the social surroundings value system together with the patient's become important.

In discussions regarding matters such as these one can sometimes hear the argument that by inflicting external values on subjects one runs the risk of tyrannizing people (the tyranny of normality). In response to such arguments it can be said that there are many tyrannies in life which have to be compared to each other. Maybe there is no a solution that is completely free from all criticism.

Psychiatric patients lack many things that are available to others. A way to handle the frustration that may result from this is to deny the need for things one cannot obtain. This is generally a natural and fruitful way to cope with frustration and disappointment. But there is a difference between needs in the sense of *wishes* and needs in the sense of *fundamental needs*. We can be denied Russian caviar and claim that it is nothing that we value, but we cannot refrain from food all together. In the same way many psychiatric patients would prefer to withdraw into a isolation; if the care system capitulates to this the patient would be abandoned to the tyranny of mental and social isolation as well.

BIBLIOGRAPHY

[1]   Allardt, E.: 1978, *Att Ha, Att Älska, Att Vara — Om välfärd i Norden*. Argos Förlag AB, Lund, Sweden.

[2]   Andrews, F. M., and Withey, S. B.: 1974, 'Developing measures of perceived life quality: Results from several national surveys'. *Social Indicators Research, 1*, 1-26.

[3]   Andrews, F. M.: 1974, 'Social indicators of perceived life quality'. *Social Indicators Research, 1*, 229-99.

[4]   Bergner, M., Bobbitt, R. A., Carter, W. B., and Gilson, B.S.: 1981, 'The Sickness Impact Profile; development and final revision of a health status measure', *Medical Care, 19*, 787-805.

[5]   Blau, T.H.: 1977, 'Quality of Life, Social Indicators and Criteria of Change', *Professional Psychology*, November, 464-73.

[6]   Bradburn, N.M.: 1969, *The Structure of Psychological Well-Being*, Chicago.

[7]   Bubholz, M.M., Eicher, J.B., Evers, S. J., and Sontag, M. S.: 1980, 'A human Ecological Approach to Quality of Life: Conceptual Framework and Results of a Preliminary Study', *Social Indicators Research, 7*, 103-36.

[8]   Campbell, A., Coverse, Ph. E., and Rogers, W. L.: 1976, *The Quality of American Life*, Russel Sage Foundation, New York.

[9]   Cantril, H.: 1965, *The Pattern of Human Concern*, Rutgers University Press, New Brunswick, New Jersey.

[10]   Diener, E.: 1984, 'Subjective well-being', *Psychological Bulletin, 95*, 542-75.

[11]   Flanagan, J.C.: 1978, 'A Research Approach to Improving Our Quality of Life', *American Psychologist*, 33, Feb., 138-47.

[12]   Gurin, G., Verhoff, J. and Feld, S.: 1969, *Americans View their Mental Health*, Basic Books, New York.

[13]   Jahoda, M.: 1958, *Current Concepts of Positive Mental Health*, Basic Books, New York.

[14]   Kajandi, M., Brattlöv, L., & Söderlind, A.: 1983, *Livskvalitet: Beräkningar av ett livskvalitetsinstruments reliabilitet*, Projekt Öppen Psykiatri, Institutionen för Psykiatri, Ulleråker, Uppsala.

[15]   Kajandi, M.: *Livskvalitet, En litteraturstudie av livskvalitet som ett beteendevetenskapligt begrepp samt ett förslag till definition*. Institutionen för psykiatri, Ulleråker, Uppsala.

[16]   Kazdin, A. E.: 'Artefact, Bias, and Complexity of Assessment: The ABC's of Reliability'. *Journal of Applied Behavior Analysis, vol. 10*, 141-50.

[17]   Kebbon, L., Nilsson, A. C., Sonnander, K. & Hjärpe, J.: 1981, *Normaliseringens kvalitet och gränser*, Projekt Mental Retardation, Institutionen för Psykiatri, Ulleråker, Uppsala.

[18]   Magnusson, D. and Öhman, A.: 1987, *Psychopathology; An interactional perspective*, Academic Press Inc., Orlando, Florida.

[19]   Michalos, A.C.: 1985, 'Multiple Discrepancies Theory (MDT)', *Social Indicators Research, 16*, 347-413.

[20]   Murrell, S.A., and Norris, F.N.: 1983, 'Quality of Life as a Criterion for Need Assessment and Community Psychology', *Journal of Community Psychology, 11*, April, 88-97.

[21]   Næss, S.: 1979, *Livskvalitet — om à ha det godt i byen og på landet*. INAS Rapport 79:2, Oslo.

[22]   Nirje, B.: 1969, 'The normalization principle and its human management implications', in Kugel, R. & Wolfensberger W. (eds.): *Changing patterns in residential services for the mentally retarded*, Presidents Committee on Mental Retardation, Washington DC.

[23]   Pörn, I: 1984, 'An Equilibrium Model of Health', in Nordenfelt, L., and Lindahl, B.I.B (eds): *Health, Disease, and Causal Explanations in Medicine*, D. Reidel Publishing Company, Dordrecht.

NOTES ON CONTRIBUTORS

Anton Aggernaes, M.D., Ph.D. (psychology), Chief physician, Department of psychiatry, Bispebjerg Hospital Copenhagen, Denmark.

Gary Albrecht, B.A., M.A., Ph.D., Professor, School of Public Health, University of Illinois, Chicago, USA.

Stefan Björk, Ph.D., Senior Research Fellow, The Swedish Institute of Health Economics, Lund, Sweden.

Mike Bury, B.A., M.Sc., Professor of Sociology, Department of Social Policy and Social Science, Royal Holloway, University of London, U.K.

Paolo Cattorini, D. Phil., M.D., Coordinator of the Department of Medical Humanities, the Scientific Institute H San Raffaele, Milan and Associate Professor of Bioethics, University of Florence, Italy.

Anne M. Fagot-Largeault, Ph.D., M.D., Professor of Philosophy, Department of Philosophy, University of Paris-X, and Consultant in Psychiatry, Henri Mondor Hospital, Assistance Publique de Paris, France.

Ray Fitzpatrick, B.A., M.Sc.,Ph.D., Fellow, Nuffield College, University Lecturer, Department of Public Health and Primary Care, University of Oxford, U.K.

Madis Kajandi, Licensed Psychotherapist, Research Psychologist, Department of Psychiatry, Uppsala University, Sweden.

Klemens Kappel, M.A., University of Copenhagen, Denmark.

Per-Erik Liss, Ph.D., Research Fellow, Department of Health and Society, Linköping University, Sweden.

Roberto Mordacci, D. Phil., Research Assistant, Department of Medical Humanities, The Scientific Institute H San Raffaele, Milan, Italy.

Torbjørn Moum, Ph.D., Professor, Department of Behavioural Sciences in Medicine, University of Oslo, Norway.

Bert Musschenga, Ph.D., Special Professor in Social Ethics, Interdisciplinary Centre for the Study of Science, Society and Religion, Vrije University of Amsterdam, The Netherlands.

Lennart Nordenfelt, Ph.D., Professor, Department of Health and Society, Linköping University, Sweden.

Siri Naess, Ph.D., Research Fellow, Institute of Applied Social Research, Oslo, Norway.

Erik Ostenfeld, M.Litt., Ph.D., Associate Professor of Classics, University of Aarhus, Denmark.

Pontus Roos, B.A., Senior Research Fellow, The Swedish Institute of Health Economics, Lund, Sweden.

Peter Sandøe, Ph.D., Senior Research Fellow, Department of Philosophy, University of Copenhagen, Denmark.

INDEX

Philosophy and Medicine

23. E.E. Shelp (ed.): *Sexuality and Medicine.*
    Vol. II: Ethical Viewpoints in Transition. 1987
                            ISBN 1-55608-013-1; Pb 1-55608-016-6
24. R.C. McMillan, H. Tristram Engelhardt, Jr., and S.F. Spicker (eds.):
    Euthanasia and the Newborn. Conflicts Regarding Saving Lives. 1987
                            ISBN 90-277-2299-4; Pb 1-55608-039-5
25. S.F. Spicker, S.R. Ingman and I.R. Lawson (eds.): *Ethical Dimensions of
    Geriatric Care.* Value Conflicts for the 21th Century. 1987
                            ISBN 1-55608-027-1
26. L. Nordenfelt: *On the Nature of Health.* An Action- Theoretic Approach. 1987
                            ISBN 1-55608-032-8
27. S.F. Spicker, W.B. Bondeson and H. Tristram Engelhardt, Jr. (eds.): *The
    Contraceptive Ethos.* Reproductive Rights and Responsibilities. 1987
                            ISBN 1-55608-035-2
28. S.F. Spicker, I. Alon, A. de Vries and H. Tristram Engelhardt, Jr. (eds.): *The
    Use of Human Beings in Research.* With Special Reference to Clinical Trials.
    1988   ISBN 1-55608-043-3
29. N.M.P. King, L.R. Churchill and A.W. Cross (eds.): *The Physician as Captain
    of the Ship.* A Critical Reappraisal. 1988   ISBN 1-55608-044-1
30. H.-M. Sass and R.U. Massey (eds.): *Health Care Systems.* Moral Conflicts in
    European and American Public Policy. 1988   ISBN 1-55608-045-X
31. R.M. Zaner (ed.): *Death: Beyond Whole-Brain Criteria.* 1988
                            ISBN 1-55608-053-0
32. B.A. Brody (ed.): *Moral Theory and Moral Judgments in Medical Ethics.* 1988
                            ISBN 1-55608-060-3
33. L.M. Kopelman and J.C. Moskop (eds.): *Children and Health Care.* Moral and
    Social Issues. 1989   ISBN 1-55608-078-6
34. E.D. Pellegrino, J.P. Langan and J. Collins Harvey (eds.): *Catholic Perspec-
    tives on Medical Morals.* Foundational Issues. 1989   ISBN 1-55608-083-2
35. B.A. Brody (ed.): *Suicide and Euthanasia.* Historical and Contemporary
    Themes. 1989   ISBN 0-7923-0106-4
36. H.A.M.J. ten Have, G.K. Kimsma and S.F. Spicker (eds.): *The Growth of
    Medical Knowledge.* 1990   ISBN 0-7923-0736-4
37. I. Löwy (ed.): *The Polish School of Philosophy of Medicine.* From Tytus
    Chałubiński (1820–1889) to Ludwik Fleck (1896–1961). 1990
                            ISBN 0-7923-0958-8
38. T.J. Bole III and W.B. Bondeson: *Rights to Health Care.* 1991
                            ISBN 0-7923-1137-X
39. M.A.G. Cutter and E.E. Shelp (eds.): *Competency.* A Study of Informal
    Competency Determinations in Primary Care. 1991   ISBN 0-7923-1304-6
40. J.L. Peset and D. Gracia (eds.): *The Ethics of Diagnosis.* 1992
                            ISBN 0-7923-1544-8

Philosophy and Medicine

41. K.W. Wildes, S.J., F. Abel, S.J. and J.C. Harvey (eds.): *Birth, Suffering, and Death.* Catholic Perspectives at the Edges of Life. 1992
ISBN 0-7923-1547-2; Pb 0-7923-2545-1

42. S.K. Toombs: *The Meaning of Illness.* A Phenomenological Account of the Different Perspectives of Physician and Patient. 1992
ISBN 0-7923-1570-7; Pb 0-7923-2443-9

43. D. Leder (ed.): *The Body in Medical Thought and Practice.* 1992
ISBN 0-7923-1657-6

44. C. Delkeskamp-Hayes and M.A.G. Cutter (eds.): *Science, Technology, and the Art of Medicine.* European-American Dialogues. 1993   ISBN 0-7923-1869-2

45. R. Baker, D. Porter and R. Porter (eds.): *The Codification of Medical Morality.* Historical and Philosophical Studies of the Formalization of Western Medical Morality in the Eighteenth and Nineteenth Centuries, Volume One: Medical Ethics and Etiquette in the Eighteenth Century. 1993   ISBN 0-7923-1921-4

46. K. Bayertz (ed.): *The Concept of Moral Consensus.* The Case of Technological Interventions into Human Reproduction. 1994   ISBN 0-7923-2615-6

47. L. Nordenfelt (ed.): *Concepts and Measurement of Quality of Life in Health Care.* 1994   ISBN 0-7923-2824-8

KLUWER ACADEMIC PUBLISHERS – DORDRECHT / BOSTON / LONDON